Renewable Energy: Power for a Sustainable Future

Renewable Energy: Power for a Sustainable Future

Edited by
Maisie Walter

Larsen & Keller
www.larsen-keller.com

Renewable Energy: Power for a Sustainable Future
Edited by Maisie Walter
ISBN: 978-1-63549-249-1 (Hardback)

© 2017 Larsen & Keller

▤ Larsen & Keller

Published by Larsen and Keller Education,
5 Penn Plaza,
19th Floor,
New York, NY 10001, USA

Cataloging-in-Publication Data

Renewable energy : power for a sustainable future / edited by Maisie Walter.
    p. cm.
Includes bibliographical references and index.
ISBN 978-1-63549-249-1
1. Renewable energy sources. 2. Power resources. 3. Renewable energy sources--Environmental aspects.
4. Geothermal resources. I. Walter, Maisie.
TJ808 .R45 2017
621.042--dc23

The publisher's policy is to use permanent paper from mills that operate a sustainable forestry policy. Furthermore, the publisher ensures that the text paper and cover boards used have met acceptable environmental accreditation standards.

Printed and bound in the United States of America.

For more information regarding Larsen and Keller Education and its products, please visit the publisher's website www.larsen-keller.com

Table of Contents

Preface

This book attempts to understand the multiple branches that fall under the discipline of renewable energy and how such concepts have practical applications. It discusses in detail the importance and application of renewable energy. It is a compilation of chapters that discuss the most vital concepts of this field. Renewable energy refers to the energy produced by the resources which are natural and renewable like sun, water, wind, geothermal heat etc. Renewable energy is an alternative to non-renewable energy sources like petroleum, fossil fuels, etc. Such selected concepts that redefine this subject area have been presented in this text. The various subfields along with technological progress that have future implications are glanced at in it. The textbook, with its diverse topics encompasses the main concern related to renewable energy which are efficient and affordable technology and responsible energy practices to the environment. Those with an interest in the field of renewable energy would find this textbook helpful.

A foreword of all chapters of the book is provided below:

Chapter 1 - The energy collected from resources such as sunlight, rain, wind, tides, and geothermal heat is defined as renewable energy. These are sources of energy that are easily replenished. The following content is an overview of the subject matter incorporating all the major aspects of renewable energy.; **Chapter 2 -** Wind energy is the kinetic energy of air in motion. This chapter explains to the reader about devices that convert the wind's kinetic energy into electrical power. Wind turbine, wind turbine design, windmill, sail and environmental impact of wind power are some of the technologies that are explained in this section.; **Chapter 3 -** Hydropower is power derived from the energy produced by falling water or fast running water. The electricity produced by hydropower is hydroelectricity. It accounts for 70% of renewable energy. The following content will provide an integrated understanding of hydropower.; **Chapter 4 -** Solar energy is light and heat from the sun that is utilized using a range of technology, such as solar energy, solar thermal energy, solar power and concentrated solar power. Solar energy is an important source of renewable energy and can be broadly characterized as either passive solar or active solar.; **Chapter 5 -** The thermal energy generated and stored in the earth is defined as geothermal energy. To have a better understanding on geothermal energy, this chapter elucidates concepts such as geothermal gradient, geothermal power and binary cycle. Geothermal energy is an emerging field of study, the following chapter will not only provide an overview, it will also delve deep into the variegated topics related to it.

At the end, I would like to thank all the people associated with this book devoting their precious time and providing their valuable contributions to this book. I would also like to express my gratitude to my fellow colleagues who encouraged me throughout the process.

Editor

Introduction of Renewable Energy

The energy collected from resources such as sunlight, rain, wind, tides, and geothermal heat is defined as renewable energy. These are sources of energy that are easily replenished. The following content is an overview of the subject matter incorporating all the major aspects of renewable energy.

Renewable energy is generally defined as energy that is collected from resources which are naturally replenished on a human timescale, such as sunlight, wind, rain, tides, waves, and geothermal heat.Renewable energy often provides energy in four important areas: electricity generation, air and water heating/cooling, transportation, and rural (off-grid) energy services.

Wind, solar, and hydroelectricity are three emerging renewable sources of energy

Based on REN21's 2016 report, renewables contributed 19.2% to humans' global energy consumption and 23.7% to their generation of electricity in 2014 and 2015, respectively. This energy consumption is divided as 8.9% coming from traditional biomass, 4.2% as heat energy (modern biomass, geothermal and solar heat), 3.9% hydro electricity and 2.2% is electricity from wind, solar, geothermal, and biomass. Worldwide investments in renewable technologies amounted to more than US$286 billion in 2015, with countries like China and the United States heavily investing in wind, hydro, solar and biofuels. Globally, there are an estimated 7.7 million jobs associated with the renewable energy industries, with solar photovoltaics being the largest renewable employer.

Renewable energy resources exist over wide geographical areas, in contrast to other energy sources, which are concentrated in a limited number of countries. Rapid deployment of renewable energy and energy efficiency is resulting in significant energy security, climate change mitigation, and economic benefits. The results of a recent review of the literature concluded that as greenhouse gas (GHG) emitters begin to be held liable for damages resulting from GHG emissions resulting in climate change, a high value for liability mitigation would provide powerful incentives for deployment of renewable energy technologies. In international public opinion surveys there is strong support for promoting renewable sources such as solar power and wind power. At the national level, at least 30 nations around the world already have renewable energy contributing more than 20 percent of energy supply. National renewable energy markets are projected to continue to grow strongly in the coming decade and beyond. Some places and at least two countries, Iceland and Norway generate all their electricity using renewable energy already, and many other countries have the set a goal to reach 100% renewable energy in the future. For example, in Denmark the government decided to switch the total energy supply (electricity, mobility and heating/cooling) to 100% renewable energy by 2050.

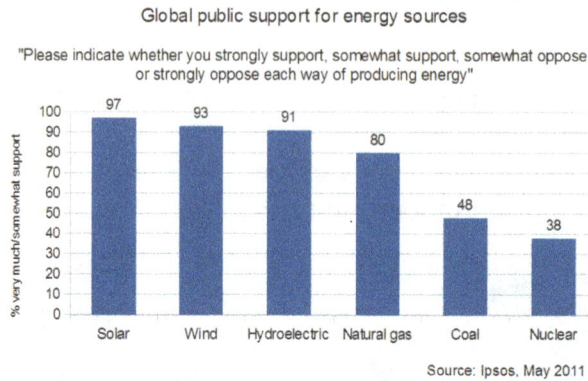

Global public support for energy sources

"Please indicate whether you strongly support, somewhat support, somewhat oppose, or strongly oppose each way of producing energy"

Source: Ipsos, May 2011

Global public support for different energy sources (2011) based on a poll by Ipsos Global @dvisor

While many renewable energy projects are large-scale, renewable technologies are also suited to rural and remote areas and developing countries, where energy is often crucial in human development. United Nations' Secretary-General Ban Ki-moon has said that renewable energy has the ability to lift the poorest nations to new levels of prosperity. As most of renewables provide electricity, renewable energy deployment is often applied in conjunction with further electrification, which has several benefits: For example, electricity can be converted to heat without losses and even reach higher temperatures than fossil fuels, can be converted into mechanical energy with high efficiency and is clean at the point of consumpion. In addition to that electrification with renewable energy is much more efficient and therefore leads to a significant reduction in primary energy requirements, because most renewables don't have a steam cycle with high losses (fossil power plants usually have losses of 40 to 65%).

Overview

Renewable

Traditional biomass	9%
Bio-heat	2.6%
Ethanol	0.34%
Biodiesel	0.15%
Biopower generation	0.25%
Hydropower	3.8%
Wind	0.39%
Solar heating/cooling	0.16%
Solar PV	0.077%
Solar CSP	0.0039%
Geothermal heat	0.061%
Geothermal electricity	0.049%
Ocean power	0.00078%

Fossil Fuel 78.4%

Petroleum

Renewable 19%

Nuclear 2.6%

Coal

Natural Gas

Total World Energy Consumption by Source (2013)

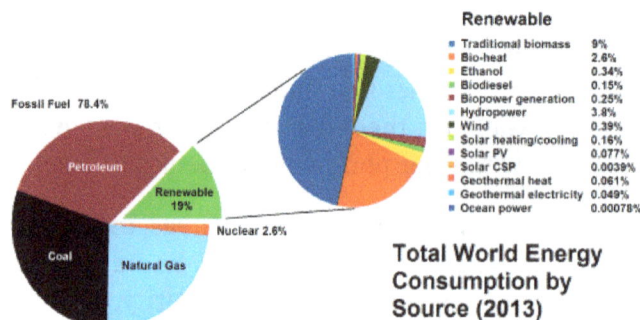

World energy consumption by source. Renewables accounted for 19% in 2012.

Renewable energy flows involve natural phenomena such as sunlight, wind, tides, plant growth, and geothermal heat, as the International Energy Agency explains:

Renewable energy is derived from natural processes that are replenished constantly. In its various forms, it derives directly from the sun, or from heat generated deep within the earth. Included in the definition is electricity and heat generated from solar, wind, ocean, hydropower, biomass, geothermal resources, and biofuels and hydrogen derived from renewable resources.

PlanetSolar, the world's largest solar-powered boat and the first ever solar electric vehicle to circumnavigate the globe (in 2012)

Renewable energy resources and significant opportunities for energy efficiency exist over wide geographical areas, in contrast to other energy sources, which are concentrated in a limited number of countries. Rapid deployment of renewable energy and energy efficiency, and technological diversification of energy sources, would result in significant energy security and economic benefits. It would also reduce environmental pollution such as air pollution caused by burning of fossil fuels and improve public health, reduce premature mortalities due to pollution and save associated health costs that amount to several hundred billion dollars annually only in the United States. Renewable energy sources, that derive their energy from the sun, either directly or indirectly, such as hydro and wind, are expected to be capable of supplying humanity energy for almost another 1 billion years, at which point the predicted increase in heat from the sun is expected to make the surface of the earth too hot for liquid water to exist.

Climate change and global warming concerns, coupled with high oil prices, peak oil, and increasing government support, are driving increasing renewable energy legislation, incentives and commercialization. New government spending, regulation and policies helped the industry weather the global financial crisis better than many other sectors. According to a 2011 projection by the International Energy Agency, solar power generators may produce most of the world's electricity within 50 years, reducing the emissions of greenhouse gases that harm the environment.

As of 2011, small solar PV systems provide electricity to a few million households, and micro-hydro configured into mini-grids serves many more. Over 44 million households use biogas made in household-scale digesters for lighting and/or cooking, and more than 166 million households rely on a new generation of more-efficient biomass cookstoves. United Nations' Secretary-General Ban Ki-moon has said that renewable energy has the ability to lift the poorest nations to new levels of prosperity. At the national level, at least 30 nations around the world already have renewable energy contributing more than 20% of energy supply. National renewable energy markets are projected to continue to grow strongly in the coming decade and beyond, and some 120 countries have various policy targets for longer-term shares of renewable energy, including a 20% target of all electricity generated for the European Union by 2020. Some countries have much higher long-term policy targets of up to 100% renewables. Outside Europe, a diverse group of 20 or more other countries target renewable energy shares in the 2020–2030 time frame that range from 10% to 50%.

Renewable energy often displaces conventional fuels in four areas: electricity generation, hot wa-ter/space heating, transportation, and rural (off-grid) energy services.

Power Generation

Renewable hydroelectric energy provides 16.3% of the worlds electricity. When hydro-electric is combined with other renewables such as wind, geothermal, solar, biomass and waste: together they make the "renewables" total, 21.7% of electricity generation worldwide as of 2013. Renewable power generators are spread across many countries, and wind power alone already provides a significant share of electricity in some areas: for example, 14% in the U.S. state of Iowa, 40% in the northern German state of Schleswig-Holstein, and 49% in Denmark. Some countries get most of their power from renewables, including Iceland (100%), Norway (98%), Brazil (86%), Austria (62%), New Zealand (65%), and Sweden (54%).

Heating

Solar water heating makes an important contribution to renewable heat in many countries, most notably in China, which now has 70% of the global total (180 GWth). Most of these systems are installed on multi-family apartment buildings and meet a portion of the hot water needs of an estimated 50–60 million households in China. Worldwide, total installed solar water heating systems meet a portion of the water heating needs of over 70 million households. The use of biomass for heating continues to grow as well. In Sweden, national use of biomass energy has surpassed that of oil. Direct geothermal for heating is also growing rapidly. The newest addition to Heating is from Geothermal Heat Pumps which provide both heating and cooling, and also flatten the electric demand curve and are thus an increasing national priority.

Transportation

Bioethanol is an alcohol made by fermentation, mostly from carbohydrates produced in sugar or starch crops such as corn, sugarcane, or sweet sorghum. Cellulosic biomass, derived from non-food sources such as trees and grasses is also being developed as a feedstock for ethanol production. Ethanol can be used as a fuel for vehicles in its pure form, but it is usually used as a gasoline additive to increase octane and improve vehicle emissions. Bioethanol is widely used in the USA and in Brazil. Biodiesel can be used as a fuel for vehicles in its pure form, but it is usually used as a diesel additive to reduce levels of particulates, carbon monoxide, and hydrocarbons from diesel-powered vehicles. Biodiesel is produced from oils or fats using transesterification and is the most common biofuel in Europe.

A bus fueled by biodiesel

A solar vehicle is an electric vehicle powered completely or significantly by direct solar energy. Usually, photovoltaic (PV) cells contained in solar panels convert the sun's energy directly into electric energy. The term "solar vehicle" usually implies that solar energy is used to power all or part of a vehicle's propulsion. Solar power may be also used to provide power for communications or controls or other auxiliary functions. Solar powered boats have mainly been limited to rivers and canals, but in 2007 an experimental 14m catamaran, the Sun21 sailed the Atlantic from Seville to Miami, and from there to New York. It was the first crossing of the Atlantic powered only by solar. Solar vehicles are not sold as practical day-to-day transportation devices at present, but are primarily demonstration vehicles and engineering exercises, often sponsored by government agencies. However, indirectly solar-charged vehicles are widespread and solar boats are available commercially.

History

Prior to the development of coal in the mid 19th century, nearly all energy used was renewable. Almost without a doubt the oldest known use of renewable energy, in the form of traditional biomass to fuel fires, dates from 790,000 years ago. Use of biomass for fire did not become commonplace until many hundreds of thousands of years later, sometime between 200,000 and 400,000 years ago. Probably the second oldest usage of renewable energy is harnessing the wind in order to drive ships over water. This practice can be traced back some 7000 years, to ships on the Nile. Moving into the time of recorded history, the primary sources of traditional renewable energy were human labor, animal power, water power, wind, in grain crushing windmills, and firewood, a traditional biomass. A graph of energy use in the United States up until 1900 shows oil and natural gas with about the same importance in 1900 as wind and solar played in 2010.

In the 1860s and '70s there were already fears that civilization would run out of fossil fuels and the need was felt for a better source. In 1873 Professor Augustine Mouchot wrote:

The time will arrive when the industry of Europe will cease to find those natural resources, so necessary for it. Petroleum springs and coal mines are not inexhaustible but are rapidly diminishing in many places. Will man, then, return to the power of water and wind? Or will he emigrate where the most powerful source of heat sends its rays to all? History will show what will come.

In 1885, Werner von Siemens, commenting on the discovery of the photovoltaic effect in the solid state, wrote:

In conclusion, I would say that however great the scientific importance of this discovery may be, its practical value will be no less obvious when we reflect that the supply of solar energy is both without limit and without cost, and that it will continue to pour down upon us for countless ages after all the coal deposits of the earth have been exhausted and forgotten.

Max Weber mentioned the end of fossil fuel in the concluding paragraphs of his Die protestant-ische Ethik und der Geist des Kapitalismus, published in 1905.

Development of solar engines continued until the outbreak of World War I. The importance of solar energy was recognized in a 1911 *Scientific American* article: "in the far distant future, natural fuels having been exhausted [solar power] will remain as the only means of existence of the human race".

The theory of peak oil was published in 1956. In the 1970s environmentalists promoted the de-velopment of renewable energy both as a replacement for the eventual depletion of oil, as well as for an escape from dependence on oil, and the first electricity generating wind turbines appeared. Solar had long been used for heating and cooling, but solar panels were too costly to build solar farms until 1980.

The IEA 2014 World Energy Outlook projects a growth of renewable energy supply from 1,700 gigawatts in 2014 to 4,550 gigawatts in 2040. Fossil fuels received about $550 billion in subsidies in 2013, compared to $120 billion for all renewable energies.

Mainstream Technologies

Wind Power

Airflows can be used to run wind turbines. Modern utility-scale wind turbines range from around 600 kW to 5 MW of rated power, although turbines with rated output of 1.5–3 MW have become the most common for commercial use; the power available from the wind is a function of the cube of the wind speed, so as wind speed increases, power output increases up to the maximum output for the par-ticular turbine. Areas where winds are stronger and more constant, such as offshore and high altitude sites, are preferred locations for wind farms. Typically full load hours of wind turbines vary between 16 and 57 percent annually, but might be higher in particularly favorable offshore sites.

The 845 MW Shepherds Flat Wind Farm near Arlington, Oregon, USA

Globally, the long-term technical potential of wind energy is believed to be five times total current global energy production, or 40 times current electricity demand, assuming all practical barriers needed were overcome. This would require wind turbines to be installed over large areas, particularly in areas of higher wind resources, such as offshore. As offshore wind speeds average ~90% greater than that of land, so offshore resources can contribute substantially more energy than land stationed turbines. In 2014 global wind generation was 706 terawatt-hours or 3% of the worlds total electricity.

Hydropower

In 2015 hydropower generated 16.6% of the worlds total electricity and 70% of all renewable electricity. Since water is about 800 times denser than air, even a slow flowing stream of water, or moderate sea swell, can yield considerable amounts of energy. There are many forms of water energy:Historically hydroelectric power came from constructing large hydroelectric dams and reservoirs, which are still popular in third world countries. The largest of which is the Three Gorges Dam(2003) in China and the Itaipu Dam(1984) built by Brazil and Paraguay.

The Three Gorges Dam on the Yangtze River in China

- Small hydro systems are hydroelectric power installations that typically produce up to 50 MW of power. They are often used on small rivers or as a low impact development on larger rivers. China is the largest producer of hydroelectricity in the world and has more than 45,000 small hydro installations.

- Run-of-the-river hydroelectricity plants derive kinetic energy from rivers without the creation of a large reservoir. This style of generation may still produce a large amount of electricity, such as the Chief Joseph Dam on the Columbia river in the United States.

Hydropower is produced in 150 countries, with the Asia-Pacific region generating 32 percent of global hydropower in 2010. For counties having the largest percentage of electricity from renewables, the top 50 are primarily hydroelectric. China is the largest hydroelectricity producer, with 721 terawatt-hours of production in 2010, representing around 17 percent of domestic electricity use. There are now three hydroelectricity stations larger than 10 GW: the Three Gorges Dam in China, Itaipu Dam across the Brazil/Paraguay border, and Guri Dam in Venezuela.

Wave power, which captures the energy of ocean surface waves, and tidal power, converting the energy of tides, are two forms of hydropower with future potential; however, they are not yet widely employed commercially. A demonstration project operated by the Ocean Renewable Power Company on the coast of Maine, and connected to the grid, harnesses tidal power from the Bay of Fundy, location of world's highest tidal flow. Ocean thermal energy conversion, which uses the temperature difference between cooler deep and warmer surface waters, has currently no economic feasibility.

Solar Energy

Solar energy, radiant light and heat from the sun, is harnessed using a range of ever-evolving technologies such as solar heating, photovoltaics, concentrated solar power (CSP), concentrator photovoltaics (CPV), solar architecture and artificial photosynthesis. Solar technologies are broadly characterized as either passive solar or active solar depending on the way they capture, convert and distribute solar energy. Passive solar techniques include orienting a building to the Sun, selecting materials with favorable thermal mass or light dispersing properties, and designing spaces that naturally circulate air. Active solar technologies encompass solar thermal energy, using solar collectors for heating, and solar power, converting sunlight into electricity either directly using photovoltaics (PV), or indirectly using concentrated solar power (CSP).

Satellite image of the 550-megawatt Topaz Solar Farm in California, USA

A photovoltaic system converts light into electrical direct current (DC) by taking advantage of the photoelectric effect. Solar PV has turned into a multi-billion, fast-growing industry, continues to improve its cost-effectiveness, and has the most potential of any renewable technologies together with CSP. Concentrated solar power (CSP) systems use lenses or mirrors and tracking systems to focus a large area of sunlight into a small beam. Commercial concentrated solar power plants were first developed in the 1980s. CSP-Stirling has by far the highest efficiency among all solar energy technologies.

In 2011, the International Energy Agency said that "the development of affordable, inexhaustible and clean solar energy technologies will have huge longer-term benefits. It will increase countries' energy security through reliance on an indigenous, inexhaustible and mostly import-independent resource, enhance sustainability, reduce pollution, lower the costs of mitigating climate change, and keep fossil fuel prices lower than otherwise. These advantages are global. Hence the additional costs of the incentives for early deployment should be considered learning investments; they must be wisely spent and need to be widely shared". In 2014 global solar generation was 186 terawatt-hours, slightly less than 1% of the worlds total grid electricity.

Geothermal Energy

High Temperature Geothermal energy is from thermal energy generated and stored in the Earth. Thermal energy is the energy that determines the temperature of matter. Earth's geothermal energy originates from the original formation of the planet and from radioactive decay of minerals (in currently uncertain but possibly roughly equal proportions). The geothermal gradient, which is the difference in temperature between the core of the planet and its surface, drives a continuous conduction of thermal energy in the form of heat from the core to the surface. The adjective *geothermal* originates from the Greek roots *geo*, meaning earth, and *thermos*, meaning heat.

Steam rising from the Nesjavellir Geothermal Power Station in Iceland

The heat that is used for geothermal energy can be from deep within the Earth, all the way down to Earth's core – 4,000 miles (6,400 km) down. At the core, temperatures may reach over 9,000 °F (5,000 °C). Heat conducts from the core to surrounding rock. Extremely high temperature and pressure cause some rock to melt, which is commonly known as magma. Magma convects upward since it is lighter than the solid rock. This magma then heats rock and water in the crust, sometimes up to 700 °F (371 °C).

From hot springs, geothermal energy has been used for bathing since Paleolithic times and for space heating since ancient Roman times, but it is now better known for electricity generation

Low Temperature Geothermal refers to the use of the outer crust of the earth as a Thermal Battery to facilitate Renewable thermal energy for heating and cooling buildings, and other refrigeration and industrial uses. In this form of Geothermal, a Geothermal Heat Pump and Ground-coupled heat exchanger are used together to move heat energy into the earth (for cooling) and out of the earth (for heating) on a varying seasonal basis. Low temperature Geothermal (generally referred to as "GHP") is an increasingly important renewable technology because it both reduces total annual energy loads associated with heating and cooling, and it also flattens the electric demand curve eliminating the extreme summer and winter peak electric supply requirements. Thus Low Temperature Geothermal/GHP is becoming an increasing national priority with multiple tax credit support and focus as part of the ongoing movement toward Net Zero Energy. New York City has even just passed a law to require GHP anytime is shown to be economical with 20 year financing including the Socialized Cost of Carbon.

Bio Energy

Biomass is biological material derived from living, or recently living organisms. It most often refers to plants or plant-derived materials which are specifically called lignocellulosic biomass. As an energy source, biomass can either be used directly via combustion to produce heat, or indirectly after converting it to various forms of biofuel. Conversion of biomass to biofuel can be achieved by different methods which are broadly classified into: *thermal*, *chemical*, and *biochemical* methods. Wood remains the largest biomass energy source today; examples include forest residues – such as dead trees, branches and tree stumps –, yard clippings, wood chips and even municipal solid waste. In the second sense, biomass includes plant or animal matter that can be converted into fibers or other industrial chemicals, including biofuels. Industrial biomass can be grown from numerous types of plants, including miscanthus, switchgrass, hemp, corn, poplar, willow, sorghum, sugarcane, bamboo, and a variety of tree species, ranging from eucalyptus to oil palm (palm oil).

Sugarcane plantation to produce ethanol in Brazil

Plant energy is produced by crops specifically grown for use as fuel that offer high biomass output per hectare with low input energy. Some examples of these plants are wheat, which typically yield 7.5–8 tonnes of grain per hectare, and straw, which typically yield 3.5–5 tonnes per hectare in the UK. The grain can be used for liquid transportation fuels while the straw can be burned to produce heat or electricity. Plant biomass can also be degraded from cellulose to glucose through a series of chemical treatments, and the resulting sugar can then be used as a first generation biofuel.

A CHP power station using wood to supply 30,000 households in France

Biomass can be converted to other usable forms of energy like methane gas or transportation fuels like ethanol and biodiesel. Rotting garbage, and agricultural and human waste, all release methane gas – also called landfill gas or biogas. Crops, such as corn and sugarcane, can be fermented to produce the transportation fuel, ethanol. Biodiesel, another transportation fuel, can be produced from left-over food products like vegetable oils and animal fats. Also, biomass to liquids (BTLs) and cellulosic ethanol are still under research. There is a great deal of research involving algal fuel or algae-derived biomass due to the fact that it's a non-food resource and can be produced at rates 5 to 10 times those of other types of land-based agriculture, such as corn and soy. Once harvested, it can be fermented to produce biofuels such as ethanol, butanol, and methane, as well as biodiesel and hydrogen. The biomass used for electricity generation varies by region. Forest by-products, such as wood residues, are common in the United States. Agricultural waste is common in Mauritius (sugar cane residue) and Southeast Asia (rice husks). Animal husbandry residues, such as poultry litter, are common in the United Kingdom.

Biofuels include a wide range of fuels which are derived from biomass. The term covers solid, liquid, and gaseous fuels. Liquid biofuels include bioalcohols, such as bioethanol, and oils, such as biodiesel. Gaseous biofuels include biogas, landfill gas and synthetic gas. Bioethanol is an alcohol made by fermenting the sugar components of plant materials and it is made mostly from sugar and starch crops. These include maize, sugarcane and, more recently, sweet sorghum. The latter crop is particularly suitable for growing in dryland conditions, and is being investigated by International Crops Research Institute for the Semi-Arid Tropics for its potential to provide fuel, along with food and animal feed, in arid parts of Asia and Africa.

With advanced technology being developed, cellulosic biomass, such as trees and grasses, are also used as feedstocks for ethanol production. Ethanol can be used as a fuel for vehicles in its pure form, but it is usually used as a gasoline additive to increase octane and improve vehicle emissions. Bioethanol is widely used in the United States and in Brazil. The energy costs for producing bio-ethanol are almost equal to, the energy yields from bio-ethanol. However, according to the European Environment Agency, biofuels do not address global warming concerns. Biodiesel is made from vegetable oils, animal fats or recycled greases. It can be used as a fuel for vehicles in its pure form, or more commonly as a diesel additive to reduce levels of particulates, carbon monoxide, and hydrocarbons from diesel-powered vehicles. Biodiesel is produced from oils or fats using transesterification and is the most common biofuel in Europe. Biofuels provided 2.7% of the world's transport fuel in 2010.

Biomass, biogas and biofuels are burned to produce heat/power and in doing so harm the environment. Pollutants such as sulphurous oxides (SO_x), nitrous oxides (NO_x), and particulate matter (PM) are produced from the combustion of biomass; the World Health Organisation estimates that 7 million premature deaths are caused each year by air pollution. Biomass combustion is a major contributor. The life cycle of the plants is sustainable, the lives of people less so.

Energy Storage

Energy storage is a collection of methods used to store electrical energy on an electrical power grid, or off it. Electrical energy is stored during times when production (especially from intermittent power plants such as renewable electricity sources such as wind power, tidal power, solar power) exceeds consumption, and returned to the grid when production falls below consumption. Water pumped into a hydroelectric dam is the largest form of power storage.

Market and Industry Trends

Growth of Renewables

From the end of 2004, worldwide renewable energy capacity grew at rates of 10–60% annually for many technologies. For wind power and many other renewable technologies, growth accelerated in 2009 relative to the previous four years. More wind power capacity was added during 2009 than any other renewable technology. However, grid-connected PV increased the fastest of all renewables technologies, with a 60% annual average growth rate. In 2010, renewable power constituted about a third of the newly built power generation capacities.

Projections vary, but scientists have advanced a plan to power 100% of the world's energy with wind, hydroelectric, and solar power by the year 2030.

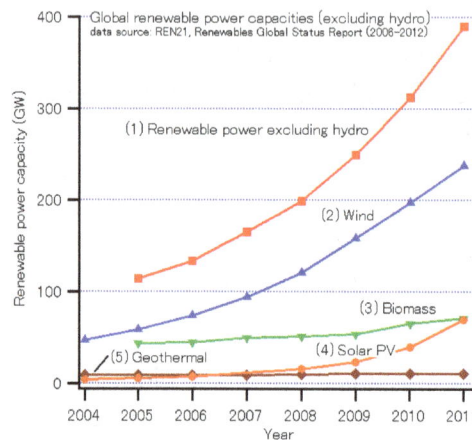

Global growth of renewables through to 2011

According to a 2011 projection by the International Energy Agency, solar power generators may produce most of the world's electricity within 50 years, reducing the emissions of greenhouse gases that harm the environment. Cedric Philibert, senior analyst in the renewable energy division at the IEA said: "Photovoltaic and solar-thermal plants may meet most of the world's demand for electricity by 2060 – and half of all energy needs – with wind, hydropower and biomass plants supplying much of the remaining generation". "Photovoltaic and concentrated solar power together can become the major source of electricity", Philibert said.

In 2014 global wind power capacity expanded 16% to 369,553 MW. Yearly wind energy production is also growing rapidly and has reached around 4% of worldwide electricity usage, 11.4% in the EU, and it is widely used in Asia, and the United States. In 2015, worldwide installed photovoltaics capacity increased to 227 gigawatts (GW), sufficient to supply 1 percent of global electricity demands. Solar thermal energy stations operate in the USA and Spain, and as of 2016, the largest of these is the 392 MW Ivanpah Solar Electric Generating System in California. The world's largest geothermal power installation is The Geysers in California, with a rated capacity of 750 MW. Brazil has one of the largest renewable energy programs in the world, involving production of ethanol fuel from sugar cane, and ethanol now provides 18% of the country's automotive fuel. Ethanol fuel is also widely available in the USA.

Selected renewable energy global indicators	2008	2009	2010	2011	2012	2013	2014	2015
Investment in new renewable capacity (annual) (10^9 USD)	182	178	237	279	256	232	270	285
Renewables power capacity (existing) (GWe)	1,140	1,230	1,320	1,360	1,470	1,578	1,712	1,849
Hydropower capacity (existing) (GWe)	885	915	945	970	990	1,018	1,055	1,064
Wind power capacity (existing) (GWe)	121	159	198	238	283	319	370	433
Solar PV capacity (grid-connected) (GWe)	16	23	40	70	100	138	177	227
Solar hot water capacity (existing) (GWth)	130	160	185	232	255	373	406	435
Ethanol production (annual) (10^9 litres)	67	76	86	86	83	87	94	98
Biodiesel production (annual) (10^9 litres)	12	17.8	18.5	21.4	22.5	26	29.7	30
Countries with policy targets for renewable energy use	79	89	98	118	138	144	164	173
Source: The Renewable Energy Policy Network for the 21st Century (REN21)–Global Status Report								

Economic Trends

Renewable energy technologies are getting cheaper, through technological change and through the benefits of mass production and market competition. A 2011 IEA report said: "A portfolio of renewable energy technologies is becoming cost-competitive in an increasingly broad range of circumstances, in some cases providing investment opportunities without the need for specific economic support," and added that "cost reductions in critical technologies, such as wind and solar, are set to continue."

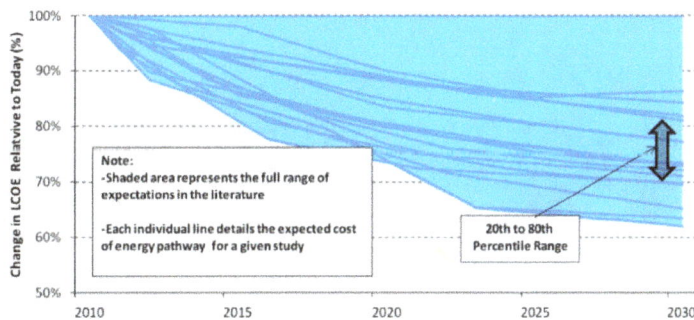

Hydro-electricity and geothermal electricity produced at favourable sites are now the cheapest way to generate electricity. Renewable energy costs continue to drop, and the levelised cost of electricity (LCOE) is declining for wind power, solar photovoltaic (PV), concentrated solar power (CSP) and some biomass technologies. Renewable energy is also the most economic solution for new grid-connected capacity in areas with good resources. As the cost of renewable power falls, the scope of economically viable applications increases. Renewable technologies are now often the most economic solution for new generating capacity. Where "oil-fired generation is the predominant power generation source (e.g. on islands, off-grid and in some countries) a lower-cost renewable solution almost always exists today". A series of studies by the US National Renewable Energy Laboratory modeled the "grid in the Western US under a number of different scenarios where intermittent renewables accounted for 33 percent of the total power." In the models, inefficiencies in cycling the fossil fuel plants to compensate for the variation in solar and wind energy resulted in an additional cost of "between $0.47 and $1.28 to each

MegaWatt hour generated"; however, the savings in the cost of the fuels saved "adds up to $7 billion, meaning the added costs are, at most, two percent of the savings."

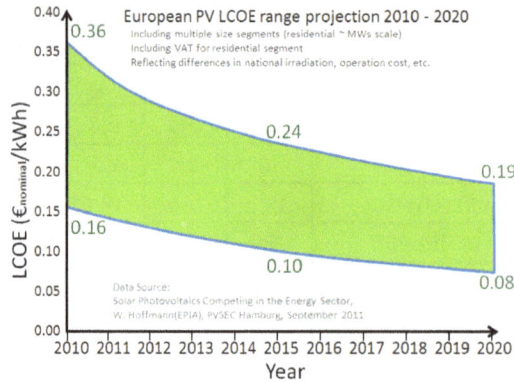

Projection of levelized cost for wind in the U.S. (left) and solar power in Europe

Hydroelectricity

Only 25% of the worlds estimated hydroelectric potential of 14,000 TWh/year has been developed, with Africa, Asia and Latin America having the greatest potential. The Three Gorges Dam in Hubei, China, has the world's largest instantaneous generating capacity (22,500 MW), with the Itaipu Dam in Brazil/Paraguay in second place (14,000 MW). The Three Gorges Dam is operated jointly with the much smaller Gezhouba Dam (3,115 MW). As of 2012, the total generating capacity of this two-dam complex is 25,615 MW. In 2008, this complex generated 98 TWh of electricity (81 TWh from the Three Gorges Dam and 17 TWh from the Gezhouba Dam), which is 3% more power in one year than the 95 TWh generated by Itaipu in 2008.

Wind Power Development

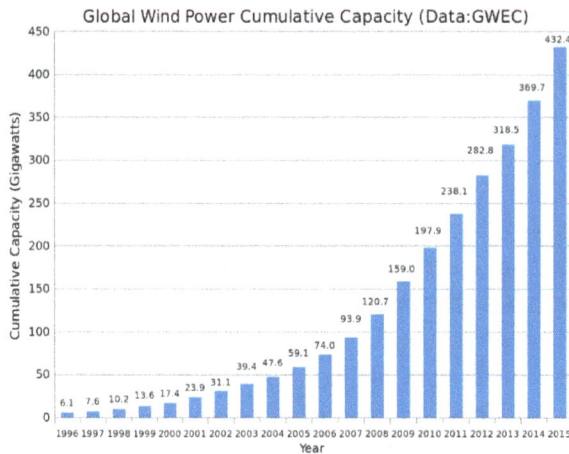

Worldwide growth of wind capacity (1996–2014)

Wind power is widely used in Europe, China, and the United States. From 2004 to 2014, worldwide installed capacity of wind power has been growing from 47 GW to 369 GW—a more than sevenfold increase within 10 years with 2014 breaking a new record in global installations (51 GW). As of the end of 2014, China, the United States and Germany combined accounted for half of total global capacity. Several other countries have achieved relatively high levels of wind power penetration, such as 21% of stationary electricity production in Denmark, 18% in Portugal, 16% in Spain, and 14% in Ireland in 2010 and have since continued to expand their installed capacity. More than 80 countries around the world are using wind power on a commercial basis.

Four offshore wind farms are in the Thames Estuary area: Kentish Flats, Gunfleet Sands, Thanet and London Array. The latter is the largest in the world as of April 2013.

- Offshore wind power

  As of 2014, offshore wind power amounted to 8,771 megawatt of global installed capacity. Although offshore capacity doubled within three years (from 4,117 MW in 2011), it accounted for only 2.3% of the total wind power capacity. The United Kingdom is the undisputed leader of offshore power with half of the world's installed capacity ahead of Denmark, Germany, Belgium and China.

- List of offshore and onshore wind farms

  As of 2012, the Alta Wind Energy Center (California, 1,020 MW) is the world's largest wind farm. The London Array (630 MW) is the largest offshore wind farm in the world. The United Kingdom is the world's leading generator of offshore wind power, followed by Denmark. There are several large offshore wind farms operational and under construction and these include Anholt (400 MW), BARD (400 MW), Clyde (548 MW), Fântânele-Cogealac (600 MW), Greater Gabbard (500 MW), Lincs (270 MW), London Array (630 MW), Lower Snake River (343 MW), Macarthur (420 MW), Shepherds Flat (845 MW), and the Sheringham Shoal (317 MW).

Solar Thermal

The United States conducted much early research in photovoltaics and concentrated solar power. The U.S. is among the top countries in the world in electricity generated by the Sun and several of the world's largest utility-scale installations are located in the desert Southwest.

The oldest solar thermal power plant in the world is the 354 megawatt (MW) SEGS thermal power plant, in California. The Ivanpah Solar Electric Generating System is a solar thermal power project in the California Mojave Desert, 40 miles (64 km) southwest of Las Vegas, with a gross capacity of 377 MW. The 280 MW Solana Generating Station is a solar power plant near Gila Bend, Arizona, about 70 miles (110 km) southwest of Phoenix, completed in 2013. When commissioned it was the largest parabolic trough plant in the world and the first U.S. solar plant with molten salt thermal energy storage.

The 377 MW Ivanpah Solar Electric Generating System with all three towers under load, Feb 2014. Taken from I-15.

The solar thermal power industry is growing rapidly with 1.3 GW under construction in 2012 and more planned. Spain is the epicenter of solar thermal power development with 873 MW under construction, and a further 271 MW under development. In the United States, 5,600 MW of solar thermal power projects have been announced. Several power plants have been constructed in the Mojave Desert, Southwestern United States. The Ivanpah Solar Power Facility being the most recent. In developing countries, three World Bank projects for integrated solar thermal/combined-cycle gas-turbine power plants in Egypt, Mexico, and Morocco have been approved.

Solar Towers of the PS10 and PS20 solar thermal plants in Spain

Photovoltaic Development

Photovoltaics (PV) uses solar cells assembled into solar panels to convert sunlight into electricity. It's a fast-growing technology doubling its worldwide installed capacity every couple of years. PV systems range from small, residential and commercial rooftop or building integrated installations, to large utility-scale photovoltaic power station. The predominant PV technology is crystalline silicon, while thin-film solar cell technology accounts for about 10 percent of global photovoltaic deployment. In recent years, PV technology has improved its electricity generating efficiency, reduced the installation cost per watt as well as its energy payback time, and has reached grid parity in at least 30 different markets by 2014. Financial institutions are predicting a second solar "gold rush" in the near future.

At the end of 2014, worldwide PV capacity reached at least 177,000 megawatts. Photovoltaics grew fastest in China, followed by Japan and the United States, while Germany remains the world's largest overall producer of photovoltaic power, contributing about 7.0 percent to the overall electricity generation. Italy meets 7.9 percent of its electricity demands with photovoltaic power—the highest share worldwide. For 2015, global cumulative capacity is forecasted to increase by more than 50 gigawatts (GW). By 2018, worldwide capacity is projected to reach as much as 430 gigawatts. This corresponds to a tripling within five years. Solar power is forecasted to become the world's largest source of electricity by 2050, with solar photovoltaics and concentrated solar power contributing 16% and 11%, respectively. This requires an increase of installed PV capacity to 4,600 GW, of which more than half is expected to be deployed in China and India.

Photovoltaic Power Stations

Commercial concentrated solar power plants were first developed in the 1980s. As the cost of solar electricity has fallen, the number of grid-connected solar PV systems has grown into the millions and utility-scale solar power stations with hundreds of megawatts are being built. Solar PV is rapidly becoming an inexpensive, low-carbon technology to harness renewable energy from the Sun.

Solar panels at the 550 MW Topaz Solar Farm

Many solar photovoltaic power stations have been built, mainly in Europe, China and the USA. The 579 MW Solar Star, in the United States, is the world's largest PV power station.

Nellis Solar Power Plant, photovoltaic power plant in Nevada, USA

Many of these plants are integrated with agriculture and some use tracking systems that follow the sun's daily path across the sky to generate more electricity than fixed-mounted systems. There are no fuel costs or emissions during operation of the power stations.

However, when it comes to renewable energy systems and PV, it is not just large systems that matter. Building-integrated photovoltaics or "onsite" PV systems use existing land and structures and generate power close to where it is consumed.

Biofuel Development

Biofuels provided 3% of the world's transport fuel in 2010. Mandates for blending biofuels exist in 31 countries at the national level and in 29 states/provinces. According to the International Energy Agency, biofuels have the potential to meet more than a quarter of world demand for transportation fuels by 2050.

Brazil produces bioethanol made from sugarcane available throughout the country. A typical gas station with dual fuel service is marked "A" for alcohol (ethanol) and "G" for gasoline.

Since the 1970s, Brazil has had an ethanol fuel program which has allowed the country to become the world's second largest producer of ethanol (after the United States) and the world's largest exporter. Brazil's ethanol fuel program uses modern equipment and cheap sugarcane as feedstock, and the residual cane-waste (bagasse) is used to produce heat and power. There are no longer light vehicles in Brazil running on pure gasoline. By the end of 2008 there were 35,000 filling stations throughout Brazil with at least one ethanol pump. Unfortunately, Operation Car Wash has seriously eroded public trust in oil companies and has implicated several high ranking Brazilian officials.

Nearly all the gasoline sold in the United States today is mixed with 10% ethanol, and motor vehicle manufacturers already produce vehicles designed to run on much higher ethanol blends. Ford, Daimler AG, and GM are among the automobile companies that sell "flexible-fuel" cars, trucks, and minivans that can use gasoline and ethanol blends ranging from pure gasoline up to 85% ethanol. By mid-2006, there were approximately 6 million ethanol compatible vehicles on U.S. roads.

Geothermal Development

Geothermal power is cost effective, reliable, sustainable, and environmentally friendly, but has historically been limited to areas near tectonic plate boundaries. Recent technological advances have expanded the range and size of viable resources, especially for applications such as home heating, opening a potential for widespread exploitation. Geothermal wells release greenhouse gases trapped deep within the earth, but these emissions are much lower per energy unit than those of fossil fuels. As a result, geothermal power has the potential to help mitigate global warming if widely deployed in place of fossil fuels.

The International Geothermal Association (IGA) has reported that 10,715 MW of geothermal power in 24 countries is online, which is expected to generate 67,246 GWh of electricity in 2010. This represents a 20% increase in geothermal power online capacity since 2005. IGA projects this will grow to 18,500 MW by 2015, due to the large number of projects presently under consideration, often in areas previously assumed to have little exploitable resource.

Geothermal plant at The Geysers, California, USA

In 2010, the United States led the world in geothermal electricity production with 3,086 MW of installed capacity from 77 power plants; the largest group of geothermal power plants in the world is located at The Geysers, a geothermal field in California. The Philippines follows the US as the second highest producer of geothermal power in the world, with 1,904 MW of capacity online; geothermal power makes up approximately 18% of the country's electricity generation.

Developing Countries

Renewable energy technology has sometimes been seen as a costly luxury item by critics, and affordable only in the affluent developed world. This erroneous view has persisted for many years, but 2015 was the first year when investment in non-hydro renewables, was higher in developing countries, with $156 billion invested, mainly in China, India, and Brazil.

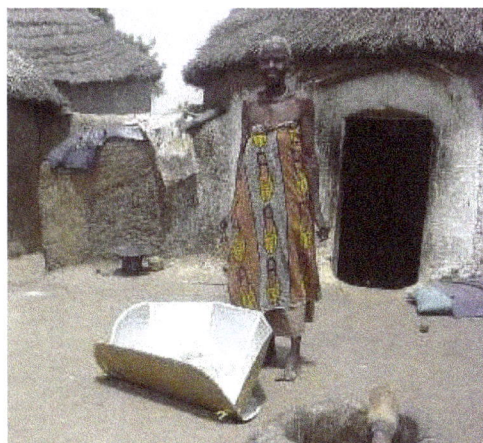

Solar cookers use sunlight as energy source for outdoor cooking

Renewable energy can be particularly suitable for developing countries. In rural and remote areas, transmission and distribution of energy generated from fossil fuels can be difficult and expensive. Producing renewable energy locally can offer a viable alternative.

Technology advances are opening up a huge new market for solar power: the approximately 1.3 billion people around the world who don't have access to grid electricity. Even though they are typically very poor, these people have to pay far more for lighting than people in rich countries because they use inefficient kerosene lamps. Solar power costs half as much as lighting with kerosene. As of 2010, an estimated 3 million households get power from small solar PV systems. Kenya is the world leader in the number of solar power systems installed per capita. More than 30,000 very small solar panels, each producing 12 to 30 watts, are sold in Kenya annually. Some Small Island Developing States (SIDS) are also turning to solar power to reduce their costs and increase their sustainability.

Micro-hydro configured into mini-grids also provide power. Over 44 million households use biogas made in household-scale digesters for lighting and/or cooking, and more than 166 million households rely on a new generation of more-efficient biomass cookstoves. Clean liquid fuel sourced from renewable feedstocks are used for cooking and lighting in energy-poor areas of the developing world. Alcohol fuels (ethanol and methanol) can be produced sustainably from non-food sugary, starchy, and cellulostic feedstocks. Project Gaia, Inc. and CleanStar Mozambique are implementing clean cooking programs with liquid ethanol stoves in Ethiopia, Kenya, Nigeria and Mozambique.

Renewable energy projects in many developing countries have demonstrated that renewable energy can directly contribute to poverty reduction by providing the energy needed for creating businesses and employment. Renewable energy technologies can also make indirect contributions to alleviating poverty by providing energy for cooking, space heating, and lighting. Renewable energy can also contribute to education, by providing electricity to schools.

Industry and Policy Trends

U.S. President Barack Obama's American Recovery and Reinvestment Act of 2009 includes more than $70 billion in direct spending and tax credits for clean energy and associated transportation programs. Leading renewable energy companies include First Solar, Gamesa, GE Energy, Hanwha Q Cells, Sharp Solar, Siemens, SunOpta, Suntech Power, and Vestas.

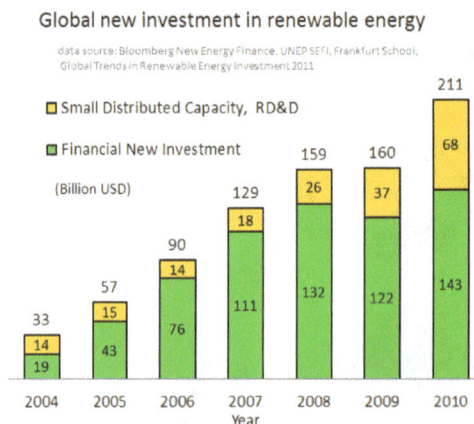

Global new investment in renewable energy

data source: Bloomberg New Energy Finance, UNEP SEFI, Frankfurt School, Global Trends in Renewable Energy Investment 2011

☐ Small Distributed Capacity, RD&D

☐ Financial New Investment

(Billion USD)

Year	2004	2005	2006	2007	2008	2009	2010
Total	33	57	90	129	159	160	211
Small Distributed Capacity, RD&D	14	15	14	18	26	37	68
Financial New Investment	19	43	76	111	132	122	143

Global New Investments in Renewable Energy

Many national, state, and local governments have also created green banks. A green bank is a quasi-public financial institution that uses public capital to leverage private investment in clean energy technologies. Green banks use a variety of financial tools to bridge market gaps that hinder the deployment of clean energy.

The military has also focused on the use of renewable fuels for military vehicles. Unlike fossil fuels, renewable fuels can be produced in any country, creating a strategic advantage. The US military has already committed itself to have 50% of its energy consumption come from alternative sources.

The International Renewable Energy Agency (IRENA) is an intergovernmental organization for promoting the adoption of renewable energy worldwide. It aims to provide concrete policy advice and facilitate capacity building and technology transfer. IRENA was formed on 26 January 2009, by 75 countries signing the charter of IRENA. As of March 2010, IRENA has 143 member states who all are considered as founding members, of which 14 have also ratified the statute.

As of 2011, 119 countries have some form of national renewable energy policy target or renewable support policy. National targets now exist in at least 98 countries. There is also a wide range of policies at state/provincial and local levels.

United Nations' Secretary-General Ban Ki-moon has said that renewable energy has the ability to lift the poorest nations to new levels of prosperity. In October 2011, he "announced the creation of a high-level group to drum up support for energy access, energy efficiency and greater use of renewable energy. The group is to be co-chaired by Kandeh Yumkella, the chair of UN Energy and director general of the UN Industrial Development Organisation, and Charles Holliday, chairman of Bank of America".

100% Renewable Energy

The incentive to use 100% renewable energy, for electricity, transport, or even total primary energy supply globally, has been motivated by global warming and other ecological as well as economic concerns. The Intergovernmental Panel on Climate Change has said that there are few fundamental technological limits to integrating a portfolio of renewable energy technologies to meet most of total global energy demand. Renewable energy use has grown much faster than even advocates anticipated. At the national level, at least 30 nations around the world already have renewable energy contributing more than 20% of energy supply. Also, Professors S. Pacala and Robert H. Socolow have developed a series of "stabilization wedges" that can allow us to maintain our quality of life while avoiding catastrophic climate change, and "renewable energy sources," in aggregate, constitute the largest number of their "wedges."

Using 100% renewable energy was first suggested in a Science paper published in 1975 by Danish physicist Bent Sørensen. It was followed by several other proposals, until in 1998 the first detailed analysis of scenarios with very high shares of renewables were published. These were followed by the first detailed 100% scenarios. In 2006 a PhD thesis was published by Czisch in which it was shown that in a 100% renewable scenario energy supply could match demand in every hour of the year in Europa and North Africa. In the same year Danish Energy professor Henrik Lund published a first paper in which he addresses the optimal combination of renewables, which was followed by several other papers on the transition to 100% renewable energy in Denmark. Since then Lund has been publishing several papers on 100% renewable energy. After 2009 publications began to rise steeply, covering 100% scenarios for countries in Europa, America, Australia and other parts of the world.

In 2011 Mark Z. Jacobson, professor of civil and environmental engineering at Stanford University, and Mark Delucchi published a study on 100% renewable global energy supply in the journal Energy Policy. They found producing all new energy with wind power, solar power, and hydropower by 2030 is feasible and existing energy supply arrangements could be replaced by 2050. Barriers to implementing the renewable energy plan are seen to be "primarily social and political, not technological or economic". They also found that energy costs with a wind, solar, water system should be similar to today's energy costs.

Similarly, in the United States, the independent National Research Council has noted that "sufficient domestic renewable resources exist to allow renewable electricity to play a significant role in future electricity generation and thus help confront issues related to climate change, energy security, and the escalation of energy costs ... Renewable energy is an attractive option because renewable resources available in the United States, taken collectively, can supply significantly greater amounts of electricity than the total current or projected domestic demand." .

The most significant barriers to the widespread implementation of large-scale renewable energy and low carbon energy strategies are primarily political and not technological. According to the 2013 *Post Carbon Pathways* report, which reviewed many international studies, the key roadblocks are: climate change denial, the fossil fuels lobby, political inaction, unsustainable energy consumption, outdated energy infrastructure, and financial constraints.

Emerging Technologies

Other renewable energy technologies are still under development, and include cellulosic ethanol, hot-dry-rock geothermal power, and marine energy. These technologies are not yet widely demonstrated or have limited commercialization. Many are on the horizon and may have potential comparable to other renewable energy technologies, but still depend on attracting sufficient attention and research, development and demonstration (RD&D) funding.

There are numerous organizations within the academic, federal, and commercial sectors conducting large scale advanced research in the field of renewable energy. This research spans several areas of focus across the renewable energy spectrum. Most of the research is targeted at improving efficiency and increasing overall energy yields. Multiple federally supported research organizations have focused on renewable energy in recent years. Two of the most prominent of these labs are Sandia National Laboratories and the National Renewable Energy Laboratory (NREL), both of which are funded by the United States Department of Energy and supported by various corporate partners. Sandia has a total budget of $2.4 billion while NREL has a budget of $375 million.

- Enhanced geothermal system

    Enhanced geothermal systems (EGS) are a new type of geothermal power technologies that do not require natural convective hydrothermal resources. The vast majority of geothermal energy within drilling reach is in dry and non-porous rock. EGS technologies "enhance" and/or create geothermal resources in this "hot dry rock (HDR)" through hydraulic stimulation. EGS and HDR technologies, like hydrothermal geothermal, are expected to be baseload resources which produce power 24 hours a day like a fossil plant. Distinct from hydrothermal, HDR and EGS may be feasible anywhere in the world, depending on the economic limits of drill depth. Good locations are over deep granite covered by a thick (3–5 km) layer of insulating sediments which slow heat loss. There are HDR and EGS systems currently being developed and tested in France, Australia, Japan, Germany, the U.S. and Switzerland. The largest EGS project in the world is a 25 megawatt demonstration plant currently being developed in the Cooper Basin, Australia. The Cooper Basin has the potential to generate 5,000–10,000 MW.

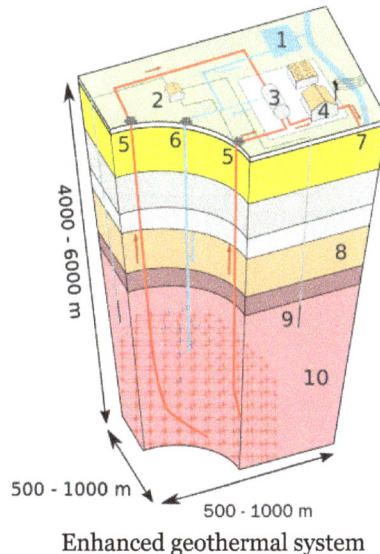

Enhanced geothermal system

- Cellulosic ethanol

    Several refineries that can process biomass and turn it into ethanol are built by companies such as Iogen, POET, and Abengoa, while other companies such as the Verenium Corporation, Novozymes, and Dyadic International are producing enzymes which could enable future commercialization. The shift from food crop feedstocks to waste residues and native grasses offers significant opportunities for a range of players, from farmers to biotechnology firms, and from project developers to investors.

- Marine energy

Rance Tidal Power Station, France

Marine energy (also sometimes referred to as ocean energy) refers to the energy carried by ocean waves, tides, salinity, and ocean temperature differences. The movement of water in the world's oceans creates a vast store of kinetic energy, or energy in motion. This energy can be harnessed to generate electricity to power homes, transport and industries. The term marine energy encompasses both wave power – power from surface waves, and tidal power – obtained from the kinetic energy of large bodies of moving water. Reverse electrodialysis (RED) is a technology for generating electricity by mixing fresh river water and salty sea water in large power cells designed for this purpose; as of 2016 it is being tested at a small scale (50 kW). Offshore wind power is not a form of marine energy, as wind power is derived from the wind, even if the wind turbines are placed over water. The oceans have a tremendous amount of energy and are close to many if not most concentrated populations. Ocean energy has the potential of providing a substantial amount of new renewable energy around the world.

#	Station	Country	Location	Capacity	Refs
1.	Sihwa Lake Tidal Power Station	South Korea	37°18′47″N 126°36′46″E37.31306°N 126.61278°E	254 MW	
2.	Rance Tidal Power Station	France	48°37′05″N 02°01′24″W48.61806°N 2.02333°W	240 MW	
3.	Annapolis Royal Generating Station	Canada	44°45′07″N 65°30′40″W44.75194°N 65.51111°W	20 MW	

- Experimental solar power

Concentrated photovoltaics (CPV) systems employ sunlight concentrated onto photovoltaic surfaces for the purpose of electricity generation. Thermoelectric, or "thermovoltaic" devices convert a temperature difference between dissimilar materials into an electric current.

- Floating solar arrays

Floating solar arrays are PV systems that float on the surface of drinking water reservoirs, quarry lakes, irrigation canals or remediation and tailing ponds. A small number of such systems exist in France, India, Japan, South Korea, the United Kingdom, Singapore and the United States. The systems are said to have advantages over photovoltaics on land. The cost of land is more expensive, and there are fewer rules and regulations for structures built on bodies of water not used for recreation. Unlike most land-based solar plants, floating arrays can be unobtrusive because they are hidden from public view. They achieve higher efficiencies than PV panels on land, because water cools the panels. The panels have a special coating to prevent rust or corrosion. In May 2008, the Far Niente Winery in Oakville, California, pioneered the world's first floatovoltaic system by installing 994 solar PV modules with a total capacity of 477 kW onto 130 pontoons and floating them on the winery's irrigation pond. Utility-scale floating PV farms are starting to be built. Kyocera will develop the world's largest, a 13.4 MW farm on the reservoir above Yamakura Dam in Chiba Prefecture using 50,000 solar panels. Salt-water resistant floating farms are also being constructed for ocean use. The largest so far announced floatovoltaic project is a 350 MW power station in the Amazon region of Brazil.

- Solar-assisted heat pump

A heat pump is a device that provides heat energy from a source of heat to a destination called a "heat sink". Heat pumps are designed to move thermal energy opposite to the direction of spontaneous heat flow by absorbing heat from a cold space and releasing it to a warmer one. A solar-assisted heat pump represents the integration of a heat pump and thermal solar panels in a single integrated system. Typically these two technologies are used separately (or only placing them in parallel) to produce hot water. In this system the solar thermal panel performs the function of the low temperature heat source and the heat produced is used to feed the heat pump's evaporator. The goal of this system is to get high COP and then produce energy in a more efficient and less expensive way.

It is possible to use any type of solar thermal panel (sheet and tubes, roll-bond, heat pipe, thermal plates) or hybrid (mono/polycrystalline, thin film) in combination with the heat pump. The use of a hybrid panel is preferable because it allows to cover a part of the electricity demand of the heat pump and reduce the power consumption and consequently the variable costs of the system.

- Artificial photosynthesis

Artificial photosynthesis uses techniques including nanotechnology to store solar electromagnetic energy in chemical bonds by splitting water to produce hydrogen and then using carbon dioxide to make methanol. Researchers in this field are striving to design molecular mimics of photosynthesis that utilize a wider region of the solar spectrum, employ catalytic systems made from abundant, inexpensive materials that are robust, readily repaired, non-toxic, stable in a variety of environmental conditions and perform more efficiently allowing a greater proportion of photon energy to end up in the storage compounds, i.e., carbohydrates (rather than building and sustaining living cells). However, prominent research faces hurdles, Sun Catalytix a MIT spin-off stopped scaling up their prototype fuel-cell in 2012, because it offers few savings over other ways to make hydrogen from sunlight.

- Algae fuels

Producing liquid fuels from oil-rich varieties of algae is an ongoing research topic. Various microalgae grown in open or closed systems are being tried including some system that can be set up in brownfield and desert lands.

- Solar aircraft

An electric aircraft is an aircraft that runs on electric motors rather than internal combustion engines, with electricity coming from fuel cells, solar cells, ultracapacitors, power beaming, or batteries.

Currently, flying manned electric aircraft are mostly experimental demonstrators, though many small unmanned aerial vehicles are powered by batteries. Electrically powered model aircraft have been flown since the 1970s, with one report in 1957. The first man-carrying electrically powered flights were made in 1973. Between 2015-2016, a manned, solar-powered plane, Solar Impulse 2, completed a circumnavigation of the Earth.

In 2016, Solar Impulse 2 was the first solar-powered aircraft to complete circumnavigation of the world.

- Solar updraft tower

The Solar updraft tower is a renewable-energy power plant for generating electricity from low temperature solar heat. Sunshine heats the air beneath a very wide greenhouse-like roofed collector structure surrounding the central base of a very tall chimney tower. The resulting convection causes a hot air updraft in the tower by the chimney effect. This airflow drives wind turbines placed in the chimney updraft or around the chimney base to produce electricity. Plans for scaled-up versions of demonstration models will allow significant power generation, and may allow development of other applications, such as water extraction or distillation, and agriculture or horticulture. A more advanced version of a similarly themed technology is the Vortex engine which aims to replace large physical chimneys with a vortex of air created by a shorter, less-expensive structure.

- Space-based solar power

For either photovoltaic or thermal systems, one option is to loft them into space, particularly Geosynchronous orbit. To be competitive with Earth-based solar power systems, the specific mass (kg/kW) times the cost to loft mass plus the cost of the parts needs to be $2400 or less. I.e., for a parts cost plus rectenna of $1100/kW, the product of the $/kg and kg/kW must be $1300/kW or less. Thus for 6.5 kg/kW, the transport cost cannot exceed $200/kg. While that will require a 100 to one reduction, SpaceX is targeting a ten to one reduction, Reaction Engines may make a 100 to one reduction possible.

- Carbon-neutral and negative fuels

Carbon-neutral fuels are synthetic fuels (including methane, gasoline, diesel fuel, jet fuel or ammonia) produced by hydrogenating waste carbon dioxide recycled from power plant flue-gas emissions, recovered from automotive exhaust gas, or derived from carbonic acid in seawater. Such fuels are considered carbon-neutral because they do not result in a net increase in atmospheric greenhouse gases. T

Carbon-neutral fuels offer relatively low cost energy storage, alleviating the problems of wind and solar variability, and they enable distribution of wind, water, and solar power through existing natural gas pipelines. Nighttime wind power is considered the most economical form of electrical power with which to synthesize fuel, because the load curve for electricity peaks sharply during the warmest hours of the day, but wind tends to blow slightly more at night than during the day, so, the price of nighttime wind power is often much less expensive than any alternative. Germany has built a 250 kilowatt synthetic methane plant which they are scaling up to 10 megawatts.

The George Olah carbon dioxide recycling plant in Grindavík, Iceland has been producing 2 million liters of methanol transportation fuel per year from flue exhaust of the Svartsengi Power Station since 2011. It has the capacity to produce 5 million liters per year.

Debate

Renewable electricity production, from sources such as wind power and solar power, is sometimes criticized for being variable or intermittent, but is not true for concentrated solar, geothermal and biofuels, that have continuity. In any case, the International Energy Agency has stated that deployment of renewable technologies usually increases the diversity of electricity sources and, through local generation, contributes to the flexibility of the system and its resistance to central shocks.

There have been "not in my back yard" (NIMBY) concerns relating to the visual and other impacts of some wind farms, with local residents sometimes fighting or blocking construction. In the USA, the Massachusetts Cape Wind project was delayed for years partly because of aesthetic concerns. However, residents in other areas have been more positive. According to a town councilor, the overwhelming majority of locals believe that the Ardrossan Wind Farm in Scotland has enhanced the area.

A recent UK Government document states that "projects are generally more likely to succeed if they have broad public support and the consent of local communities. This means giving communities both a say and a stake". In countries such as Germany and Denmark many renewable projects are owned by communities, particularly through cooperative structures, and contribute significantly to overall levels of renewable energy deployment.

The market for renewable energy technologies has continued to grow. Climate change concerns and increasing in green jobs, coupled with high oil prices, peak oil, oil wars, oil spills, promotion of electric vehicles and renewable electricity, nuclear disasters and increasing government support, are driving increasing renewable energy legislation, incentives and commercialization. New government spending, regulation and policies helped the industry weather the 2009 economic crisis better than many other sectors.

References

- Spellman, Frank R. (2013). Safe Work Practices for Green Energy Jobs (first ed.). DEStech Publications. p. 323. ISBN 978-1-60595-075-4. Retrieved 29 December 2014.

- Batrawy, Aya (9 March 2015). "Solar-powered plane takes off for flight around the world". Associated Press. Retrieved 14 March 2015.

- "The surprising history of sustainable energy". Sustainablehistory.wordpress.com. Archived from the original on Dec 24, 2014. Retrieved 1 November 2012.

- Hubbert, M. King (June 1956). "Nuclear Energy and the Fossil Fuels" (PDF). Shell Oil Company/American Petroleum Institute. Retrieved 10 November 2014.

- "2014 Outlook: Let the Second Gold Rush Begin" (PDF). Deutsche Bank Markets Research. 6 January 2014. Archived from the original on 22 November 2014. Retrieved 22 November 2014.

- "Global Market Outlook for Photovoltaics 2014-2018" (PDF). epia.org. EPIA – European Photovoltaic Industry Association. Archived from the original on 12 June 2014. Retrieved 12 June 2014.

- "Kyocera, partners announce construction of the world's largest floating solar PV Plant in Hyogo prefecture, Japan". SolarServer.com. 4 September 2014.

- Kyocera and Century Tokyo Leasing to Develop 13.4MW Floating Solar Power Plant on Reservoir in Chiba Prefecture, Japan, Kyocera, December 22, 2014

- "REN21, Renewables Global Status Report 2012". Ren21.net. Archived from the original (PDF) on 11 August 2014. Retrieved 11 August 2014.

- "Energy and environment policy case for a global project on artificial photosynthesis". Energy & Environmental Science. RSC Publishing. 6: 695. doi:10.1039/C3EE00063J. Retrieved 19 August 2013.

- Morris C & Pehnt M, German Energy Transition: Arguments for a Renewable Energy Future, Heinrich Böll Foundation, November 2012

- Simon Gourlay (12 August 2008). "Wind farms are not only beautiful, they're absolutely necessary". The Guardian. UK. Retrieved 17 January 2012.

2

Wind Energy

Wind energy is the kinetic energy of air in motion. This chapter explains to the reader about devices that convert the wind's kinetic energy into electrical power. Wind turbine, wind turbine design, windmill, sail and environmental impact of wind power are some of the technologies that are explained in this section.

Wind Power

Wind power is the use of air flow through wind turbines to mechanically power generators for electricity. Wind power, as an alternative to burning fossil fuels, is plentiful, renewable, widely distributed, clean, produces no greenhouse gas emissions during operation, consumes no water, and uses little land. The net effects on the environment are far less problematic than those of non-renewable power sources.

Wind power stations in Xinjiang, China

Wind farms consist of many individual wind turbines which are connected to the electric power transmission network. Onshore wind is an inexpensive source of electricity, competitive with or in many places cheaper than coal or gas plants. Offshore wind is steadier and stronger than on land, and offshore farms have less visual impact, but construction and maintenance costs are considerably higher. Small onshore wind farms can feed some energy into the grid or provide electricity to isolated off-grid locations.

Wind power gives variable power which is very consistent from year to year but which has significant variation over shorter time scales. It is therefore used in conjunction with other electric power sources to give a reliable supply. As the proportion of wind power in a region increases, a need to upgrade the grid, and a lowered ability to supplant conventional production can occur. Power management techniques such as having excess capacity, geographically distributed turbines, dispatchable backing sources, sufficient hydroelectric power, exporting and importing power to neighboring areas, using vehicle-to-grid strategies or reducing demand when wind production is low, can in many cases overcome these problems. In addition, weather forecasting permits the electricity network to be readied for the predictable variations in production that occur.

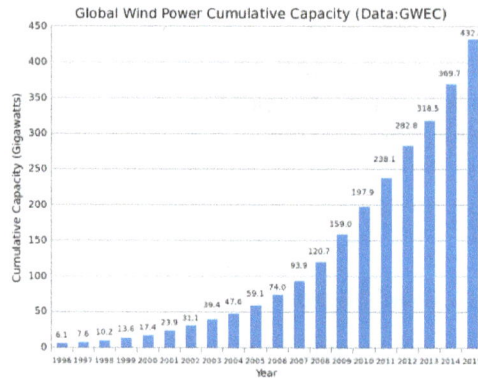

Global growth of installed capacity

As of 2015, Denmark generates 40% of its electricity from wind, and at least 83 other countries around the world are using wind power to supply their electricity grids. In 2014 global wind power capacity expanded 16% to 369,553 MW. Yearly wind energy production is also growing rapidly and has reached around 4% of worldwide electricity usage, 11.4% in the EU.

History

Wind power has been used as long as humans have put sails into the wind. For more than two millennia wind-powered machines have ground grain and pumped water. Wind power was widely available and not confined to the banks of fast-flowing streams, or later, requiring sources of fuel. Wind-powered pumps drained the polders of the Netherlands, and in arid regions such as the American mid-west or the Australian outback, wind pumps provided water for live stock and steam engines.

Charles Brush's windmill of 1888, used for generating electricity.

The first windmill used for the production of electricity was built in Scotland in July 1887 by Prof James Blyth of Anderson's College, Glasgow (the precursor of Strathclyde University). Blyth's 10 m high, cloth-sailed wind turbine was installed in the garden of his holiday cottage at Marykirk in Kincardineshire and was used to charge accumulators developed by the Frenchman Camille Alphonse Faure, to power the lighting in the cottage, thus making it the first house in the world to have its electricity supplied by wind power. Blyth offered the surplus electricity to the people of Marykirk for lighting the main street, however, they turned down the offer as they thought electricity was "the work of the devil." Although he later built a wind turbine to supply emergency power to the local Lunatic Asylum, Infirmary and Dispensary of Montrose the invention never really caught on as the technology was not considered to be economically viable.

Across the Atlantic, in Cleveland, Ohio a larger and heavily engineered machine was designed and constructed in the winter of 1887–1888 by Charles F. Brush, this was built by his engineering company at his home and operated from 1886 until 1900. The Brush wind turbine had a rotor 17 m (56 foot) in diameter and was mounted on an 18 m (60 foot) tower. Although large by today's standards, the machine was only rated at 12 kW. The connected dynamo was used either to charge a bank of batteries or to operate up to 100 incandescent light bulbs, three arc lamps, and various motors in Brush's laboratory.

With the development of electric power, wind power found new applications in lighting buildings remote from centrally-generated power. Throughout the 20th century parallel paths developed small wind stations suitable for farms or residences, and larger utility-scale wind generators that could be connected to electricity grids for remote use of power. Today wind powered generators operate in every size range between tiny stations for battery charging at isolated residences, up to near-gigawatt sized offshore wind farms that provide electricity to national electrical networks.

Wind Farms

Large onshore wind farms			
Wind farm	**Current capacity (MW)**	**Country**	**Refs**
Gansu Wind Farm	6,000	China	
Muppandal wind farm	1,500	India	
Alta (Oak Creek-Mojave)	1,320	United States	
Jaisalmer Wind Park	1,064	India	
Shepherds Flat Wind Farm	845	United States	
Roscoe Wind Farm	782	United States	
Horse Hollow Wind Energy Center	736	United States	
Capricorn Ridge Wind Farm	662	United States	
Fântânele-Cogealac Wind Farm	600	Romania	
Fowler Ridge Wind Farm	600	United States	
Whitelee Wind Farm	539	United Kingdom	

A wind farm is a group of wind turbines in the same location used for production of electricity. A large wind farm may consist of several hundred individual wind turbines distributed over an extended area, but the land between the turbines may be used for agricultural or other purposes. For example, Gansu Wind Farm, the largest wind farm in the world, has several thousand turbines. A wind farm may also be located offshore.

Almost all large wind turbines have the same design — a horizontal axis wind turbine having an upwind rotor with three blades, attached to a nacelle on top of a tall tubular tower.

In a wind farm, individual turbines are interconnected with a medium voltage (often 34.5 kV), power collection system and communications network. In general, a distance of 7D (7 × Rotor Diameter of the Wind Turbine) is set between each turbine in a fully developed wind farm. At a substation, this medium-voltage electric current is increased in voltage with a transformer for connection to the high voltage electric power transmission system.

Generator Characteristics and Stability

Induction generators, which were often used for wind power projects in the 1980s and 1990s, require reactive power for excitation so substations used in wind-power collection systems include substantial capacitor banks for power factor correction. Different types of wind turbine generators behave differently during transmission grid disturbances, so extensive modelling of the dynamic electromechanical characteristics of a new wind farm is required by transmission system operators to ensure predictable stable behaviour during system faults. In particular, induction generators cannot support the system voltage during faults, unlike steam or hydro turbine-driven synchronous generators.

Today these generators aren't used any more in modern turbines. Instead today most turbines use variable speed generators combined with partial- or full-scale power converter between the turbine generator and the collector system, which generally have more desirable properties for grid interconnection and have Low voltage ride through-capabilities. Modern concepts use either doubly fed machines with partial-scale converters or squirrel-cage induction generators or synchronous generators (both permanently and electrically excited) with full scale converters.

Transmission systems operators will supply a wind farm developer with a grid code to specify the requirements for interconnection to the transmission grid. This will include power factor, constancy of frequency and dynamic behaviour of the wind farm turbines during a system fault.

Offshore Wind Power

Offshore wind power refers to the construction of wind farms in large bodies of water to generate electricity. These installations can utilize the more frequent and powerful winds that are available in these locations and have less aesthetic impact on the landscape than land based projects. However, the construction and the maintenance costs are considerably higher.

Siemens and Vestas are the leading turbine suppliers for offshore wind power. DONG Energy, Vattenfall and E.ON are the leading offshore operators. As of October 2010, 3.16 GW of offshore wind power capacity was operational, mainly in Northern Europe. According to BTM Consult, more than 16 GW of additional capacity will be installed before the end of 2014 and the UK and Germany will become the two leading markets. Offshore wind power capacity is expected to reach a total of 75 GW worldwide by 2020, with significant contributions from China and the US.

The world's second full-scale floating wind turbine (and first to be installed without the use of heavy-lift vessels), WindFloat, operating at rated capacity (2 MW) approximately 5 km offshore of Póvoa de Varzim, Portugal

In 2012, 1,662 turbines at 55 offshore wind farms in 10 European countries produced 18 TWh, enough to power almost five million households. As of August 2013 the London Array in the United Kingdom is the largest offshore wind farm in the world at 630 MW. This is followed by Gwynt y Môr (576 MW), also in the UK.

World's largest offshore wind farms				
Wind farm	**Capacity (MW)**	**Country**	**Turbines and model**	**Commissioned**
London Array	630	🇬🇧 United Kingdom	175 × Siemens SWT-3.6	2012
Gwynt y Môr	576	🇬🇧 United Kingdom	160 × Siemens SWT-3.6 107	2015
Greater Gabbard	504	🇬🇧 United Kingdom	140 × Siemens SWT-3.6	2012
Anholt	400	🇩🇰 Denmark	111 × Siemens SWT-3.6–120	2013
BARD Offshore 1	400	🇩🇪 Germany	80 BARD 5.0 turbines	2013

Collection and Transmission Network

In a wind farm, individual turbines are interconnected with a medium voltage (usually 34.5 kV) power collection system and communications network. At a substation, this medium-voltage electric current is increased in voltage with a transformer for connection to the high voltage electric power transmission system.

A transmission line is required to bring the generated power to (often remote) markets. For an off-shore station this may require a submarine cable. Construction of a new high-voltage line may be too costly for the wind resource alone, but wind sites may take advantage of lines installed for conventionally fueled generation.

One of the biggest current challenges to wind power grid integration in the United States is the necessity of developing new transmission lines to carry power from wind farms, usually in remote lowly populated states in the middle of the country due to availability of wind, to high load locations, usually on the coasts where population density is higher. The current transmission lines in remote locations were not designed for the transport of large amounts of energy. As transmission lines become longer the losses associated with power transmission increase, as modes of losses at lower lengths are exacerbated and new modes of losses are no longer negligible as the length is increased, making it harder to transport large loads over large distances. However, resistance from state and local governments makes it difficult to construct new transmission lines. Multi state power transmission projects are discouraged by states with cheap electricity rates for fear that exporting their cheap power will lead to increased rates. A 2005 energy law gave the Energy Department authority to approve transmission projects states refused to act on, but after an attempt to use this authority, the Senate declared the department was being overly aggressive in doing so. Another problem is that wind companies find out after the fact that the transmission capacity of a new farm is below the generation capacity, largely because federal utility rules to encourage renewable energy installation allow feeder lines to meet only minimum standards. These are important issues that need to be solved, as when the transmission capacity does not meet the generation capacity, wind farms are forced to produce below their full potential or stop running all together, in a process known as curtailment. While this leads to potential renewable generation left untapped, it prevents possible grid overload or risk to reliable service.

Wind Power Capacity and Production

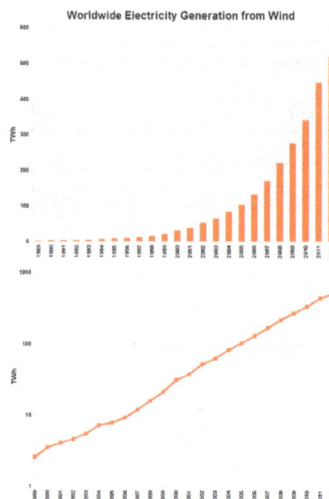

Worldwide wind generation up to 2012 (Source EIA, January 2015).

Worldwide there are now over two hundred thousand wind turbines operating, with a total nameplate capacity of 432,000 MW as of end 2015. The European Union alone passed some 100,000 MW nameplate capacity in September 2012, while the United States surpassed 75,000 MW in 2015 and China's grid connected capacity passed 145,000 MW in 2015.

World wind generation capacity more than quadrupled between 2000 and 2006, doubling about every three years. The United States pioneered wind farms and led the world in installed capacity in the 1980s and into the 1990s. In 1997 installed capacity in Germany surpassed the U.S. and led until once again overtaken by the U.S. in 2008. China has been rapidly expanding its wind installations in the late 2000s and passed the U.S. in 2010 to become the world leader. As of 2011, 83 countries around the world were using wind power on a commercial basis.

Wind power capacity has expanded rapidly to 336 GW in June 2014, and wind energy production was around 4% of total worldwide electricity usage, and growing rapidly. The actual amount of electricity that wind is able to generate is calculated by multiplying the nameplate capacity by the capacity factor, which varies according to equipment and location. Estimates of the capacity factors for wind installations are in the range of 35% to 44%.

Europe accounted for 48% of the world total wind power generation capacity in 2009. In 2010, Spain became Europe's leading producer of wind energy, achieving 42,976 GWh. Germany held the top spot in Europe in terms of installed capacity, with a total of 27,215 MW as of 31 December 2010. In 2015 wind power constituted 15.6% of all installed power generation capacity in the EU and it generates around 11.4% of its power.

Worldwide cumulative capacity, 2014

Top windpower electricity producing countries in 2012 (TWh)		
Country	**Windpower Production**	**% of World Total**
United States	140.9	26.4
China	118.1	22.1
Spain	49.1	9.2
Germany	46.0	8.6
India	30.0	5.6
United Kingdom	19.6	3.7
France	14.9	2.8
Italy	13.4	2.5
Canada	11.8	2.2
Denmark	10.3	1.9
(rest of world)	80.2	15.0
World Total	**534.3 TWh**	**100%**
Source:*Observ'ER – Electricity Production From Wind Sources [2012]*		

Growth Trends

The wind power industry set new records in 2014- more than 50GW of new capacity was installed. Another record breaking year occurred in 2015, with 22% annual market growth resulting in the 60 GW mark being passed. In 2015, close to half of all new wind power was added outside of the traditional markets in Europe and North America. This was largely from new construction in China and India. Global Wind Energy Council (GWEC) figures show that 2015 recorded an increase of installed capacity of more than 63 GW, taking the total installed wind energy capacity to 432.9 GW, up from 74 GW in 2006. In terms of economic value, the wind energy sector has become one of the important players in the energy markets, with the total investments reaching US$329bn (€296.6bn), an increase of 4% over 2014.

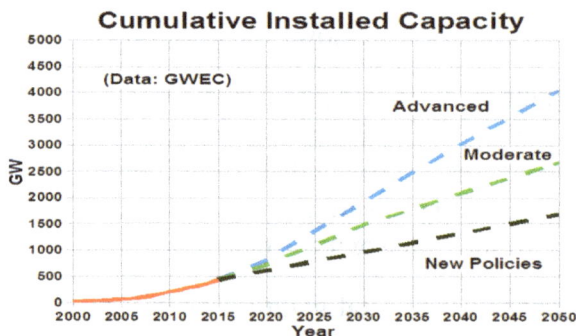

Worldwide installed wind power capacity forecast (Source: Global Wind Energy Council)

Although the wind power industry was affected by the global financial crisis in 2009 and 2010, GWEC predicts that the installed capacity of wind power will be 792.1 GW by the end of 2020 and 4,042 GW by end of 2050. The increased commissioning of wind power is being accompanied by record low prices for forthcoming renewable electricity. In some cases, wind onshore is already the cheapest electricity generation option and costs are continuing to decline. The contracted prices for wind onshore for the next few years are now as low as 30 USD/MWh.

In the EU in 2015, 44% of all new generating capacity was wind power; while in the same period net fossil fuel power capacity decreased.

Capacity Factor

Since wind speed is not constant, a wind farm's annual energy production is never as much as the sum of the generator nameplate ratings multiplied by the total hours in a year. The ratio of actual productivity in a year to this theoretical maximum is called the capacity factor. Typical capacity factors are 15–50%; values at the upper end of the range are achieved in favourable sites and are due to wind turbine design improvements.

Online data is available for some locations, and the capacity factor can be calculated from the yearly output. For example, the German nationwide average wind power capacity factor over all of 2012 was just under 17.5% (45867 GW·h/yr / (29.9 GW × 24 × 366) = 0.1746), and the capacity factor for Scottish wind farms averaged 24% between 2008 and 2010.

Unlike fueled generating plants, the capacity factor is affected by several parameters, including the variability of the wind at the site and the size of the generator relative to the turbine's swept area. A small generator would be cheaper and achieve a higher capacity factor but would produce less electricity (and thus less profit) in high winds. Conversely, a large generator would cost more but generate little extra power and, depending on the type, may stall out at low wind speed. Thus an optimum capacity factor of around 40–50% would be aimed for.

A 2008 study released by the U.S. Department of Energy noted that the capacity factor of new wind installations was increasing as the technology improves, and projected further improvements for future capacity factors. In 2010, the department estimated the capacity factor of new wind turbines in 2010 to be 45%. The annual average capacity factor for wind generation in the US has varied between 29.8% and 34.0% during the period 2010–2015.

Penetration

Wind energy penetration is the fraction of energy produced by wind compared with the total generation. The wind power penetration in world electricity generation in 2015 was 3.5%.

Country	Penetration
Denmark (2015)	42.1%
Portugal (2013)	23%
Spain (2011)	16%
Ireland (2012)	16%
United Kingdom (2014)	9.3%
Germany (2011)	8%
United States (2013)	4.5%

There is no generally accepted maximum level of wind penetration. The limit for a particular grid will depend on the existing generating plants, pricing mechanisms, capacity for energy storage, demand management and other factors. An interconnected electricity grid will already include reserve generating and transmission capacity to allow for equipment failures. This reserve capacity can also serve to compensate for the varying power generation produced by wind stations. Studies have indicated that 20% of the total annual electrical energy consumption may be incorporated with minimal difficulty. These studies have been for locations with geographically dispersed wind farms, some degree of dispatchable energy or hydropower with storage capacity, demand management, and interconnected to a large grid area enabling the export of electricity when needed. Beyond the 20% level, there are few technical limits, but the economic implications become more significant. Electrical utilities continue to study the effects of large scale penetration of wind generation on system stability and economics.

A wind energy penetration figure can be specified for different durations of time, but is often quoted annually. To obtain 100% from wind annually requires substantial long term storage or substantial interconnection to other systems which may already have substantial storage. On a monthly, weekly, daily, or hourly basis—or less—wind might supply as much as or more than 100% of current use, with the rest stored or exported. Seasonal industry might then take advantage of high wind and low usage times such as at night when wind output can exceed normal demand. Such industry might include production of silicon, aluminum, steel, or of natural gas, and hydrogen, and using future long term storage to facilitate 100% energy from variable renewable energy. Homes can also be programmed to accept extra electricity on demand, for example by remotely turning up water heater thermostats.

In Australia, the state of South Australia generates around half of the nation's wind power capacity. By the end of 2011 wind power in South Australia, championed by Premier (and Climate

Change Minister) Mike Rann, reached 26% of the State's electricity generation, edging out coal for the first time. At this stage South Australia, with only 7.2% of Australia's population, had 54% of Australia's installed capacity.

Variability

Electricity generated from wind power can be highly variable at several different timescales: hourly, daily, or seasonally. Annual variation also exists, but is not as significant. Because instantaneous electrical generation and consumption must remain in balance to maintain grid stability, this variability can present substantial challenges to incorporating large amounts of wind power into a grid system. Intermittency and the non-dispatchable nature of wind energy production can raise costs for regulation, incremental operating reserve, and (at high penetration levels) could require an increase in the already existing energy demand management, load shedding, storage solutions or system interconnection with HVDC cables.

Windmills are typically installed in favourable windy locations. In the image, wind power generators in Spain, near an Osborne bull.

Fluctuations in load and allowance for failure of large fossil-fuel generating units require reserve capacity that can also compensate for variability of wind generation.

Increase in system operation costs, Euros per MWh, for 10% & 20% wind share		
Country	10%	20%
Germany	2.5	3.2
Denmark	0.4	0.8
Finland	0.3	1.5
Norway	0.1	0.3
Sweden	0.3	0.7

Wind power is variable, and during low wind periods it must be replaced by other power sources. Transmission networks presently cope with outages of other generation plants and daily changes in electrical demand, but the variability of intermittent power sources such as wind power, are unlike those of conventional power generation plants which, when scheduled to be operating, may be able to deliver their nameplate capacity around 95% of the time.

Presently, grid systems with large wind penetration require a small increase in the frequency of usage of natural gas spinning reserve power plants to prevent a loss of electricity in the event that conditions are not favorable for power production from the wind. At lower wind power grid penetration, this is less of an issue.

GE has installed a prototype wind turbine with onboard battery similar to that of an electric car, equivalent of 1 minute of production. Despite the small capacity, it is enough to guarantee that power output complies with forecast for 15 minutes, as the battery is used to eliminate the difference rather than provide full output. The increased predictability can be used to take wind power penetration from 20 to 30 or 40 per cent. The battery cost can be retrieved by selling burst power on demand and reducing backup needs from gas plants.

A report on Denmark's wind power noted that their wind power network provided less than 1% of average demand on 54 days during the year 2002. Wind power advocates argue that these periods of low wind can be dealt with by simply restarting existing power stations that have been held in readiness, or interlinking with HVDC. Electrical grids with slow-responding thermal power plants and without ties to networks with hydroelectric generation may have to limit the use of wind power. According to a 2007 Stanford University study published in the *Journal of Applied Meteorology and Climatology*, interconnecting ten or more wind farms can allow an average of 33% of the total energy produced (i.e. about 8% of total nameplate capacity) to be used as reliable, baseload electric power which can be relied on to handle peak loads, as long as minimum criteria are met for wind speed and turbine height.

Conversely, on particularly windy days, even with penetration levels of 16%, wind power generation can surpass all other electricity sources in a country. In Spain, in the early hours of 16 April 2012 wind power production reached the highest percentage of electricity production till then, at 60.46% of the total demand. In Denmark, which had power market penetration of 30% in 2013, over 90 hours, wind power generated 100% of the country's power, peaking at 122% of the country's demand at 2 am on 28 October.

A 2006 International Energy Agency forum presented costs for managing intermittency as a function of wind-energy's share of total capacity for several countries, as shown in the table on the right. Three reports on the wind variability in the UK issued in 2009, generally agree that variability of wind needs to be taken into account, but it does not make the grid unmanageable. The additional costs, which are modest, can be quantified.

The combination of diversifying variable renewables by type and location, forecasting their variation, and integrating them with dispatchable renewables, flexible fueled generators, and demand response can create a power system that has the potential to meet power supply needs reliably. Integrating ever-higher levels of renewables is being successfully demonstrated in the real world:

In 2009, eight American and three European authorities, writing in the leading electrical engineers' professional journal, didn't find "a credible and firm technical limit to the amount of wind energy that can be accommodated by electricity grids". In fact, not one of more than 200 international studies, nor official studies for the eastern and western U.S. regions, nor the International Energy Agency, has found major costs or technical barriers to reliably integrating up to 30% variable renewable supplies into the grid, and in some studies much more. – *Reinventing Fire*

Solar power tends to be complementary to wind. On daily to weekly timescales, high pressure areas tend to bring clear skies and low surface winds, whereas low pressure areas tend to be windier and cloudier. On seasonal timescales, solar energy peaks in summer, whereas in many areas wind energy is lower in summer and higher in winter. Thus the intermittencies of wind and solar power tend to cancel each other somewhat. In 2007 the Institute for Solar Energy Supply Technology of the University of Kassel pilot-tested a combined power plant linking solar, wind, biogas and hydrostorage to provide load-following power around the clock and throughout the year, entirely from renewable sources.

Predictability

Wind power forecasting methods are used, but predictability of any particular wind farm is low for short-term operation. For any particular generator there is an 80% chance that wind output will change less than 10% in an hour and a 40% chance that it will change 10% or more in 5 hours.

However, studies by Graham Sinden (2009) suggest that, in practice, the variations in thousands of wind turbines, spread out over several different sites and wind regimes, are smoothed. As the distance between sites increases, the correlation between wind speeds measured at those sites, decreases.

Thus, while the output from a single turbine can vary greatly and rapidly as local wind speeds vary, as more turbines are connected over larger and larger areas the average power output becomes less variable and more predictable.

Wind power hardly ever suffers major technical failures, since failures of individual wind turbines have hardly any effect on overall power, so that the distributed wind power is reliable and predictable, whereas conventional generators, while far less variable, can suffer major unpredictable outages.

Energy Storage

The Sir Adam Beck Generating Complex at Niagara Falls, Canada, includes a large pumped-storage hydroelectricity reservoir. During hours of low electrical demand excess electrical grid power is used to pump water up into the reservoir, which then provides an extra 174 MW of electricity during periods of peak demand.

Typically, conventional hydroelectricity complements wind power very well. When the wind is blowing strongly, nearby hydroelectric stations can temporarily hold back their water. When the wind drops they can, provided they have the generation capacity, rapidly increase production to compensate.

This gives a very even overall power supply and virtually no loss of energy and uses no more water.

Alternatively, where a suitable head of water is not available, pumped-storage hydroelectricity or other forms of grid energy storage such as compressed air energy storage and thermal energy storage can store energy developed by high-wind periods and release it when needed. The type of storage needed depends on the wind penetration level – low penetration requires daily storage, and high penetration requires both short and long term storage – as long as a month or more. Stored energy increases the economic value of wind energy since it can be shifted to displace higher cost generation during peak demand periods. The potential revenue from this arbitrage can offset the cost and losses of storage; the cost of storage may add 25% to the cost of any wind energy stored but it is not envisaged that this would apply to a large proportion of wind energy generated. For example, in the UK, the 1.7 GW Dinorwig pumped-storage plant evens out electrical demand peaks, and allows base-load suppliers to run their plants more efficiently. Although pumped-storage power systems are only about 75% efficient, and have high installation costs, their low running costs and ability to reduce the required electrical base-load can save both fuel and total electrical generation costs.

In particular geographic regions, peak wind speeds may not coincide with peak demand for electrical power. In the U.S. states of California and Texas, for example, hot days in summer may have low wind speed and high electrical demand due to the use of air conditioning. Some utilities subsidize the purchase of geothermal heat pumps by their customers, to reduce electricity demand during the summer months by making air conditioning up to 70% more efficient; widespread adoption of this technology would better match electricity demand to wind availability in areas with hot summers and low summer winds. A possible future option may be to interconnect widely dispersed geographic areas with an IIVDC "super grid". In the U.S. it is estimated that to upgrade the transmission system to take in planned or potential renewables would cost at least $60 billion, while the society value of added windpower would be more than that cost.

Germany has an installed capacity of wind and solar that can exceed daily demand, and has been exporting peak power to neighboring countries, with exports which amounted to some 14.7 billion kilowatt hours in 2012. A more practical solution is the installation of thirty days storage capacity able to supply 80% of demand, which will become necessary when most of Europe's energy is obtained from wind power and solar power. Just as the EU requires member countries to maintain 90 days strategic reserves of oil it can be expected that countries will provide electricity storage, instead of expecting to use their neighbors for net metering.

Capacity Credit, Fuel Savings and Energy Payback

The capacity credit of wind is estimated by determining the capacity of conventional plants displaced by wind power, whilst maintaining the same degree of system security. According to the American Wind Energy Association, production of wind power in the United States in 2015 avoided consumption of 73 billion gallons of water and reduced CO_2 emissions by 132 million metric tons, while providing $7.3 billion in public health savings.

The energy needed to build a wind farm divided into the total output over its life, Energy Return on Energy Invested, of wind power varies but averages about 20–25. Thus, the energy payback time is typically around one year.

Economics

Wind turbines reached grid parity (the point at which the cost of wind power matches traditional sources) in some areas of Europe in the mid-2000s, and in the US around the same time. Falling prices continue to drive the levelized cost down and it has been suggested that it has reached general grid parity in Europe in 2010, and will reach the same point in the US around 2016 due to an expected reduction in capital costs of about 12%.

Electricity Cost and Trends

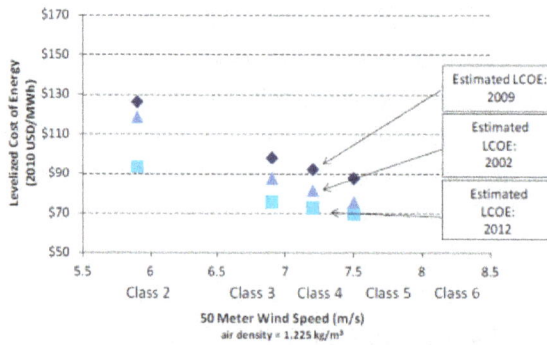

Estimated cost per MWh for wind power in Denmark

Wind power is capital intensive, but has no fuel costs. The price of wind power is therefore much more stable than the volatile prices of fossil fuel sources. The marginal cost of wind energy once a station is constructed is usually less than 1-cent per kW·h.

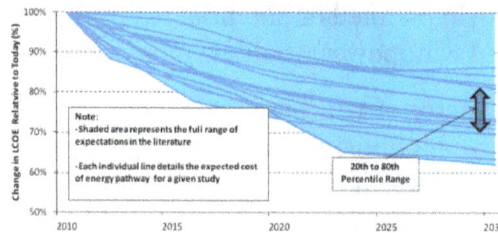

The National Renewable Energy Laboratory projects that the levelized cost of wind power in the U.S. will decline about 25% from 2012 to 2030.

However, the estimated average cost per unit of electricity must incorporate the cost of construction of the turbine and transmission facilities, borrowed funds, return to investors (including cost of risk), estimated annual production, and other components, averaged over the projected useful life of the equipment, which may be in excess of twenty years. Energy cost estimates are highly dependent on these assumptions so published cost figures can differ substantially. In 2004, wind energy cost a fifth of what it did in the 1980s, and some expected that downward trend to continue as larger multi-megawatt turbines were mass-produced. In 2012 capital costs for wind turbines were substantially lower than 2008–2010 but still above 2002 levels. A 2011 report from the

American Wind Energy Association stated, "Wind's costs have dropped over the past two years, in the range of 5 to 6 cents per kilowatt-hour recently.... about 2 cents cheaper than coal-fired electricity, and more projects were financed through debt arrangements than tax equity structures last year.... winning more mainstream acceptance from Wall Street's banks.... Equipment makers can also deliver products in the same year that they are ordered instead of waiting up to three years as was the case in previous cycles.... 5,600 MW of new installed capacity is under construction in the United States, more than double the number at this point in 2010. Thirty-five percent of all new power generation built in the United States since 2005 has come from wind, more than new gas and coal plants combined, as power providers are increasingly enticed to wind as a convenient hedge against unpredictable commodity price moves."

A turbine blade convoy passing through Edenfield in the U.K. (2008). Even longer two-piece blades are now manufactured, and then assembled on-site to reduce difficulties in transportation.

A British Wind Energy Association report gives an average generation cost of onshore wind power of around 3.2 pence (between US 5 and 6 cents) per kW·h (2005). Cost per unit of energy produced was estimated in 2006 to be 5 to 6 percent above the cost of new generating capacity in the US for coal and natural gas: wind cost was estimated at $55.80 per MW·h, coal at $53.10/MW·h and natural gas at $52.50. Similar comparative results with natural gas were obtained in a governmental study in the UK in 2011. In 2011 power from wind turbines could be already cheaper than fossil or nuclear plants; it is also expected that wind power will be the cheapest form of energy generation in the future. The presence of wind energy, even when subsidised, can reduce costs for consumers (€5 billion/yr in Germany) by reducing the marginal price, by minimising the use of expensive peaking power plants.

An 2012 EU study shows base cost of onshore wind power similar to coal, when subsidies and externalities are disregarded. Wind power has some of the lowest external costs.

In February 2013 Bloomberg New Energy Finance (BNEF) reported that the cost of generating electricity from new wind farms is cheaper than new coal or new baseload gas plants. When including the current Australian federal government carbon pricing scheme their modeling gives costs (in Australian dollars) of $80/MWh for new wind farms, $143/MWh for new coal plants and $116/MWh for new baseload gas plants. The modeling also shows that "even without a carbon price (the most efficient way to reduce economy-wide emissions) wind energy is

14% cheaper than new coal and 18% cheaper than new gas." Part of the higher costs for new coal plants is due to high financial lending costs because of "the reputational damage of emissions-intensive investments". The expense of gas fired plants is partly due to "export market" effects on local prices. Costs of production from coal fired plants built in "the 1970s and 1980s" are cheaper than renewable energy sources because of depreciation. In 2015 BNEF calculated LCOE prices per MWh energy in new powerplants (excluding carbon costs) : $85 for onshore wind ($175 for offshore), $66–75 for coal in the Americas ($82–105 in Europe), gas $80–100. A 2014 study showed unsubsidized LCOE costs between $37–81, depending on region. A 2014 US DOE report showed that in some cases power purchase agreement prices for wind power had dropped to record lows of $23.5/MWh.

The cost has reduced as wind turbine technology has improved. There are now longer and lighter wind turbine blades, improvements in turbine performance and increased power generation efficiency. Also, wind project capital and maintenance costs have continued to decline. For example, the wind industry in the USA in early 2014 were able to produce more power at lower cost by using taller wind turbines with longer blades, capturing the faster winds at higher elevations. This has opened up new opportunities and in Indiana, Michigan, and Ohio, the price of power from wind turbines built 300 feet to 400 feet above the ground can now compete with conventional fossil fuels like coal. Prices have fallen to about 4 cents per kilowatt-hour in some cases and utilities have been increasing the amount of wind energy in their portfolio, saying it is their cheapest option.

A number of initiatives are working to reduce costs of electricity from offshore wind. One example is the Carbon Trust Offshore Wind Accelerator, a joint industry project, involving nine offshore wind developers, which aims to reduce the cost of offshore wind by 10% by 2015. It has been suggested that innovation at scale could deliver 25% cost reduction in offshore wind by 2020. Henrik Stiesdal, former Chief Technical Officer at Siemens Wind Power, has stated that by 2025 energy from offshore wind will be one of the cheapest, scalable solutions in the UK, compared to other renewables and fossil fuel energy sources, if the true cost to society is factored into the cost of energy equation. He calculates the cost at that time to be 43 EUR/MWh for onshore, and 72 EUR/MWh for offshore wind.

Incentives and Community Benefits

U.S. landowners typically receive $3,000–$5,000 annual rental income per wind turbine, while farmers continue to grow crops or graze cattle up to the foot of the turbines. Shown: the Brazos Wind Farm, Texas.

The U.S. wind industry generates tens of thousands of jobs and billions of dollars of economic activity. Wind projects provide local taxes, or payments in lieu of taxes and strengthen the economy of rural communities by providing income to farmers with wind turbines on their land. Wind energy in many jurisdictions receives financial or other support to encourage its development. Wind energy benefits from subsidies in many jurisdictions, either to increase its attractiveness, or to compensate for subsidies received by other forms of production which have significant negative externalities.

Some of the 6,000 turbines in California's Altamont Pass Wind Farm aided by tax incentives during the 1980s.

In the US, wind power receives a production tax credit (PTC) of 1.5¢/kWh in 1993 dollars for each kW·h produced, for the first ten years; at 2.2 cents per kW·h in 2012, the credit was renewed on 2 January 2012, to include construction begun in 2013. A 30% tax credit can be applied instead of receiving the PTC. Another tax benefit is accelerated depreciation. Many American states also provide incentives, such as exemption from property tax, mandated purchases, and additional markets for "green credits". The Energy Improvement and Extension Act of 2008 contains extensions of credits for wind, including microturbines. Countries such as Canada and Germany also provide incentives for wind turbine construction, such as tax credits or minimum purchase prices for wind generation, with assured grid access (sometimes referred to as feed-in tariffs). These feed-in tariffs are typically set well above average electricity prices. In December 2013 U.S. Senator Lamar Alexander and other Republican senators argued that the "wind energy production tax credit should be allowed to expire at the end of 2013" and it expired 1 January 2014 for new installations.

Secondary market forces also provide incentives for businesses to use wind-generated power, even if there is a premium price for the electricity. For example, socially responsible manufacturers pay utility companies a premium that goes to subsidize and build new wind power infrastructure. Companies use wind-generated power, and in return they can claim that they are undertaking strong "green" efforts. In the US the organization Green-e monitors business compliance with these renewable energy credits.

Small-scale Wind Power

Small-scale wind power is the name given to wind generation systems with the capacity to produce up to 50 kW of electrical power. Isolated communities, that may otherwise rely on diesel generators, may use wind turbines as an alternative. Individuals may purchase these systems to reduce or eliminate their dependence on grid electricity for economic reasons, or to reduce their carbon footprint. Wind turbines have been used for household electricity generation in conjunction with battery storage over many decades in remote areas.

Recent examples of small-scale wind power projects in an urban setting can be found in New York City, where, since 2009, a number of building projects have capped their roofs with Gorlov-type helical wind turbines. Although the energy they generate is small compared to the buildings' over-all consumption, they help to reinforce the building's 'green' credentials in ways that "showing people your high-tech boiler" can not, with some of the projects also receiving the direct support of the New York State Energy Research and Development Authority.

Grid-connected domestic wind turbines may use grid energy storage, thus replacing purchased electricity with locally produced power when available. The surplus power produced by domestic microgenerators can, in some jurisdictions, be fed into the network and sold to the utility company, producing a retail credit for the microgenerators' owners to offset their energy costs.

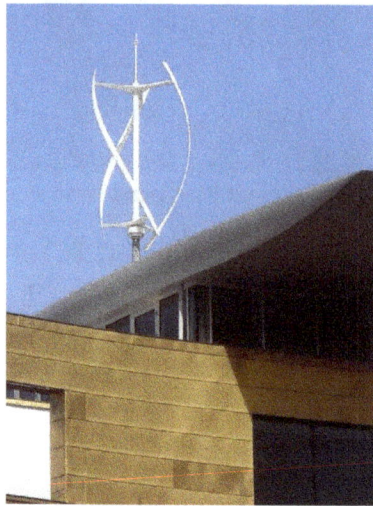

A small Quietrevolution QR5 Gorlov type vertical axis wind turbine on the roof of Colston Hall in Bristol, England. Measuring 3 m in diameter and 5 m high, it has a nameplate rating of 6.5 kW.

Off-grid system users can either adapt to intermittent power or use batteries, photovoltaic or die-sel systems to supplement the wind turbine. Equipment such as parking meters, traffic warning signs, street lighting, or wireless Internet gateways may be powered by a small wind turbine, pos-sibly combined with a photovoltaic system, that charges a small battery replacing the need for a connection to the power grid.

A Carbon Trust study into the potential of small-scale wind energy in the UK, published in 2010, found that small wind turbines could provide up to 1.5 terawatt hours (TW·h) per year of electricity (0.4% of total UK electricity consumption), saving 0.6 million tonnes of carbon dioxide (Mt CO_2) emission savings. This is based on the assumption that 10% of households would install turbines at costs competitive with grid electricity, around 12 pence (US 19 cents) a kW·h. A report prepared for the UK's government-sponsored Energy Saving Trust in 2006, found that home power generators of various kinds could provide 30 to 40% of the country's electricity needs by 2050.

Distributed generation from renewable resources is increasing as a consequence of the increased awareness of climate change. The electronic interfaces required to connect renewable generation units with the utility system can include additional functions, such as the active filtering to enhance the power quality.

Livestock grazing near a wind turbine.

The environmental impact of wind power when compared to the environmental impacts of fossil fuels, is relatively minor. According to the IPCC, in assessments of the life-cycle global warming potential of energy sources, wind turbines have a median value of between 12 and 11 (gCO_2eq/kWh) depending on whether off- or onshore turbines are being assessed. Compared with other low carbon power sources, wind turbines have some of the lowest global warming potential per unit of electrical energy generated.

While a wind farm may cover a large area of land, many land uses such as agriculture are compatible with it, as only small areas of turbine foundations and infrastructure are made unavailable for use.

There are reports of bird and bat mortality at wind turbines as there are around other artificial structures. The scale of the ecological impact may or may not be significant, depending on specific circumstances. Prevention and mitigation of wildlife fatalities, and protection of peat bogs, affect the siting and operation of wind turbines.

Wind turbines generate some noise. At a residential distance of 300 metres (980 ft) this may be around 45 dB, which is slightly louder than a refrigerator. At 1.5 km (1 mi) distance they become inaudible. There are anecdotal reports of negative health effects from noise on people who live very close to wind turbines. Peer-reviewed research has generally not supported these claims.

The United States Air Force and Navy have expressed concern that siting large windmills near bases "will negatively impact radar to the point that air traffic controllers will lose the location of aircraft."

Aesthetic aspects of wind turbines and resulting changes of the visual landscape are significant. Conflicts arise especially in scenic and heritage protected landscapes.

Politics

Central Government

Nuclear power and fossil fuels are subsidized by many governments, and wind power and other forms of renewable energy are also often subsidized. For example, a 2009 study by the Environmental Law Institute assessed the size and structure of U.S. energy subsidies over the 2002–2008 period. The study estimated that subsidies to fossil-fuel based sources amounted to approximately $72 billion over this period and subsidies to renewable fuel sources totalled $29 billion. In the United States, the federal government has paid US$74 billion for energy subsidies to support R&D for nuclear power ($50 billion) and fossil fuels ($24 billion) from 1973 to 2003. During this same time frame, renewable energy technologies and energy efficiency received a total of US$26 billion. It has been suggested that a subsidy shift would help to level the playing field and support growing energy sectors, namely solar power, wind power, and biofuels. History shows that no energy sector was developed without subsidies.

Part of the Seto Hill Windfarm in Japan.

According to the International Energy Agency (IEA) (2011), energy subsidies artificially lower the price of energy paid by consumers, raise the price received by producers or lower the cost of production. "Fossil fuels subsidies costs generally outweigh the benefits. Subsidies to renewables and low-carbon energy technologies can bring long-term economic and environmental benefits". In November 2011, an IEA report entitled *Deploying Renewables 2011* said "subsidies in green energy technologies that were not yet competitive are justified in order to give an incentive to investing into technologies with clear environmental and energy security benefits". The IEA's report disagreed with claims that renewable energy technologies are only viable through costly subsidies and not able to produce energy reliably to meet demand.

In the U.S., the wind power industry has recently increased its lobbying efforts considerably, spending about $5 million in 2009 after years of relative obscurity in Washington. By comparison, the U.S. nuclear industry alone spent over $650 million on its lobbying efforts and campaign contributions during a single ten-year period ending in 2008.

Following the 2011 Japanese nuclear accidents, Germany's federal government is working on a new plan for increasing energy efficiency and renewable energy commercialization, with a particular focus on offshore wind farms. Under the plan, large wind turbines will be erected far away from the coastlines, where the wind blows more consistently than it does on land, and where the enormous turbines won't bother the inhabitants. The plan aims to decrease Germany's dependence on energy derived from coal and nuclear power plants.

Public Opinion

Surveys of public attitudes across Europe and in many other countries show strong public support for wind power. About 80% of EU citizens support wind power. In Germany, where wind power has gained very high social acceptance, hundreds of thousands of people have invested in citizens' wind farms across the country and thousands of small and medium-sized enterprises are running successful businesses in a new sector that in 2008 employed 90,000 people and generated 8% of Germany's electricity.

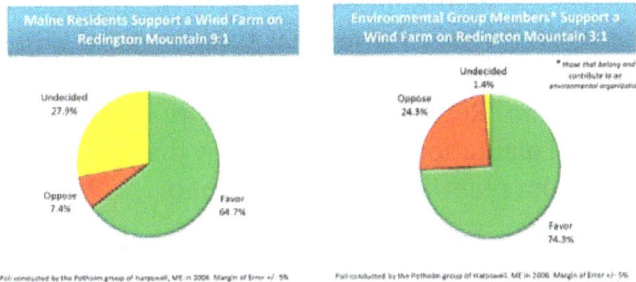

Environmental group members are both more in favor of wind power (74%) as well as more opposed (24%). Few are undecided.

Although wind power is a popular form of energy generation, the construction of wind farms is not universally welcomed, often for aesthetic reasons.

In Spain, with some exceptions, there has been little opposition to the installation of inland wind parks. However, the projects to build offshore parks have been more controversial. In particular, the proposal of building the biggest offshore wind power production facility in the world in southwestern Spain in the coast of Cádiz, on the spot of the 1805 Battle of Trafalgar has been met with strong opposition who fear for tourism and fisheries in the area, and because the area is a war grave.

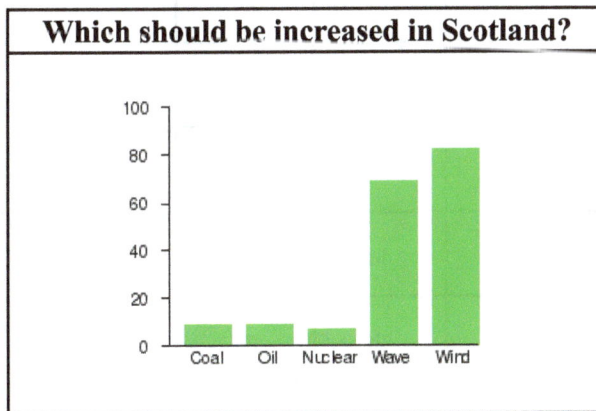

In a survey conducted by Angus Reid Strategies in October 2007, 89 per cent of respondents said that using renewable energy sources like wind or solar power was positive for Canada, because these sources were better for the environment. Only 4 per cent considered using renewable sources as negative since they can be unreliable and expensive. According to a Saint Consulting survey in April 2007, wind power was the alternative energy source most likely to gain public support for future development in Canada, with only 16% opposed to this type of energy. By contrast, 3 out of 4 Canadians opposed nuclear power developments.

A 2003 survey of residents living around Scotland's 10 existing wind farms found high levels of community acceptance and strong support for wind power, with much support from those who lived closest to the wind farms. The results of this survey support those of an earlier Scottish Executive survey 'Public attitudes to the Environment in Scotland 2002', which found that the Scottish public would prefer the majority of their electricity to come from renewables, and which rated wind power as the cleanest source of renewable energy. A survey conducted in 2005 showed that 74% of people in Scotland agree that wind farms are necessary to meet current and future energy needs. When people were asked the same question in a Scottish renewables study conducted in 2010, 78% agreed. The increase is significant as there were twice as many wind farms in 2010 as there were in 2005. The 2010 survey also showed that 52% disagreed with the statement that wind farms are "ugly and a blot on the landscape". 59% agreed that wind farms were necessary and that how they looked was unimportant. Regarding tourism, query responders consider power pylons, cell phone towers, quarries and plantations more negatively than wind farms. Scotland is planning to obtain 100% of electricity from renewable sources by 2020.

In other cases there is direct community ownership of wind farm projects. The hundreds of thousands of people who have become involved in Germany's small and medium-sized wind farms demonstrate such support there.

This 2010 Harris Poll reflects the strong support for wind power in Germany, other European countries, and the U.S.

Opinion on increase in number of wind farms Public Opinion, 2010 Harris Poll						
	U.S.	Great Britain	France	Italy	Spain	Germany
	%	%	%	%	%	%
Strongly oppose	3	6	6	2	2	4
Oppose more than favour	9	12	16	11	9	14
Favour more than oppose	37	44	44	38	37	42
Strongly favour	50	38	33	49	53	40

Community

Many wind power companies work with local communities to reduce environmental and other concerns associated with particular wind farms. In other cases there is direct community ownership of wind farm projects. Appropriate government consultation, planning and approval procedures also help to minimize environmental risks. Some may still object to wind farms but, according to The Australia Institute, their concerns should be weighed against the need to address the threats posed by climate change and the opinions of the broader community.

In America, wind projects are reported to boost local tax bases, helping to pay for schools, roads and hospitals. Wind projects also revitalize the economy of rural communities by providing steady income to farmers and other landowners.

Wind turbines such as these, in Cumbria, England, have been opposed for a number of reasons, including aesthetics, by some sectors of the population.

In the UK, both the National Trust and the Campaign to Protect Rural England have expressed concerns about the effects on the rural landscape caused by inappropriately sited wind turbines and wind farms.

Some wind farms have become tourist attractions. The Whitelee Wind Farm Visitor Centre has an exhibition room, a learning hub, a café with a viewing deck and also a shop. It is run by the Glasgow Science Centre.

In Denmark, a loss-of-value scheme gives people the right to claim compensation for loss of value of their property if it is caused by proximity to a wind turbine. The loss must be at least 1% of the property's value.

Despite this general support for the concept of wind power in the public at large, local opposition often exists and has delayed or aborted a number of projects. For example, there are concerns that some installations can negatively affect TV and radio reception and Doppler weather radar, as well as produce excessive sound and vibration levels leading to a decrease in property values. Potential broadcast-reception solutions include predictive interference modeling as a component of site selection. A study of 50,000 home sales near wind turbines found no statistical evidence that prices were affected.

A panoramic view of the United Kingdom's Whitelee Wind Farm with Lochgoin Reservoir in the foreground.

While aesthetic issues are subjective and some find wind farms pleasant and optimistic, or symbols of energy independence and local prosperity, protest groups are often formed to attempt to block new wind power sites for various reasons.

This type of opposition is often described as NIMBYism, but research carried out in 2009 found that there is little evidence to support the belief that residents only object to renewable power facilities such as wind turbines as a result of a "Not in my Back Yard" attitude.

Turbine Design

Wind turbines are devices that convert the wind's kinetic energy into electrical power. The result of over a millennium of windmill development and modern engineering, today's wind turbines are manufactured in a wide range of horizontal axis and vertical axis types. The smallest turbines are used for applications such as battery charging for auxiliary power. Slightly larger turbines can be used for making small contributions to a domestic power supply while selling unused power back to the utility supplier via the electrical grid. Arrays of large turbines, known as wind farms, have become an increasingly important source of renewable energy and are used in many countries as part of a strategy to reduce their reliance on fossil fuels.

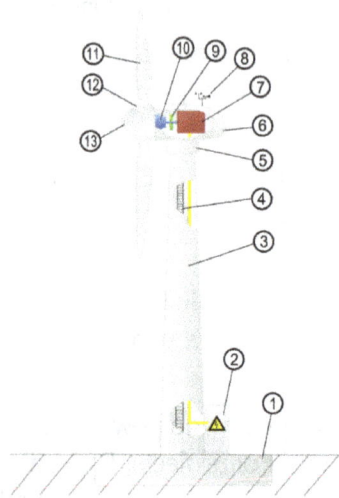

Typical wind turbine components : 1-Foundation, 2-Connection to the electric grid, 3-Tower, 4-Access ladder, 5-Wind orientation control (Yaw control), 6-Nacelle, 7-Generator, 8-Anemometer, 9-Electric or Mechanical Brake, 10-Gearbox, 11-Rotor blade, 12-Blade pitch control, 13-Rotor hub.

Typical components of a wind turbine (gearbox, rotor shaft and brake assembly) being lifted into position

Wind turbine design is the process of defining the form and specifications of a wind turbine to extract energy from the wind. A wind turbine installation consists of the necessary systems needed to capture the wind's energy, point the turbine into the wind, convert mechanical rotation into electrical power, and other systems to start, stop, and control the turbine.

In 1919 the German physicist Albert Betz showed that for a hypothetical ideal wind-energy extraction machine, the fundamental laws of conservation of mass and energy allowed no more than 16/27 (59.3%) of the kinetic energy of the wind to be captured. This Betz limit can be approached in modern turbine designs, which may reach 70 to 80% of the theoretical Betz limit.

The aerodynamics of a wind turbine are not straightforward. The air flow at the blades is not the same as the airflow far away from the turbine. The very nature of the way in which energy is extracted from the air also causes air to be deflected by the turbine. In addition the aerodynamics of a wind turbine at the rotor surface exhibit phenomena that are rarely seen in other aerodynamic fields. The shape and dimensions of the blades of the wind turbine are determined by the aerodynamic performance required to efficiently extract energy from the wind, and by the strength required to resist the forces on the blade.

In addition to the aerodynamic design of the blades, the design of a complete wind power system must also address the design of the installation's rotor hub, nacelle, tower structure, generator, controls, and foundation. Further design factors must also be considered when integrating wind turbines into electrical power grids.

Wind Energy

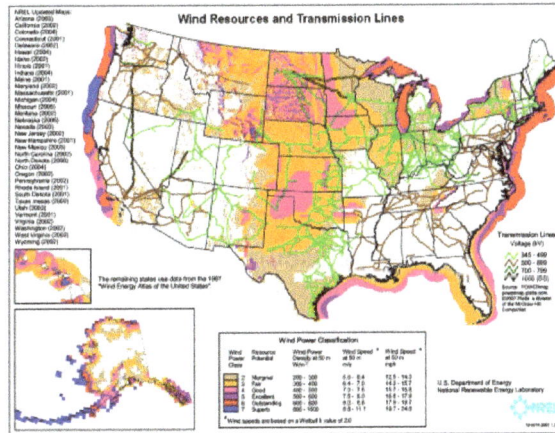

Map of available wind power for the United States. Color codes indicate wind power density class.

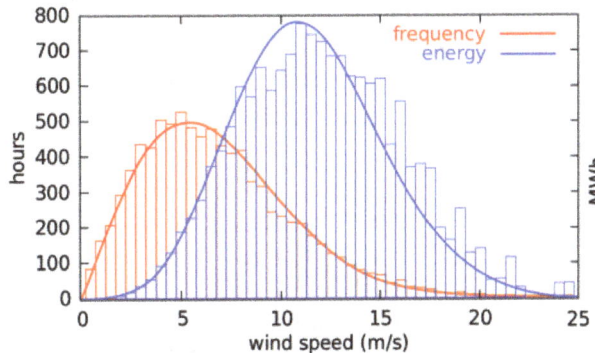

Distribution of wind speed (red) and energy (blue) for all of 2002 at the Lee Ranch facility in Colorado. The histogram shows measured data, while the curve is the Rayleigh model distribution for the same average wind speed.

Wind energy is the kinetic energy of air in motion, also called wind. Total wind energy flowing through an imaginary surface with area A during the time t is:

$$E = \frac{1}{2}mv^2 = \frac{1}{2}(Avt\rho)v^2 = \frac{1}{2}At\rho v^3,$$

where ρ is the density of air; v is the wind speed; Avt is the volume of air passing through A (which is considered perpendicular to the direction of the wind); $Avt\rho$ is therefore the mass m passing through "A". Note that $\frac{1}{2}\rho v^2$ is the kinetic energy of the moving air per unit volume.

Power is energy per unit time, so the wind power incident on A (e.g. equal to the rotor area of a wind turbine) is:

$$P = \frac{E}{t} = \frac{1}{2}A\rho v^3.$$

Wind power in an open air stream is thus *proportional* to the *third power* of the wind speed; the available power increases eightfold when the wind speed doubles. Wind turbines for grid electricity therefore need to be especially efficient at greater wind speeds.

Wind is the movement of air across the surface of the Earth, affected by areas of high pressure and of low pressure. The global wind kinetic energy averaged approximately 1.50 MJ/m² over the period from 1979 to 2010, 1.31 MJ/m² in the Northern Hemisphere with 1.70 MJ/m² in the Southern Hemisphere. The atmosphere acts as a thermal engine, absorbing heat at higher temperatures, releasing heat at lower temperatures. The process is responsible for production of wind kinetic energy at a rate of 2.46 W/m² sustaining thus the circulation of the atmosphere against frictional dis-sipation.

The total amount of economically extractable power available from the wind is considerably more than present human power use from all sources. Axel Kleidon of the Max Planck Institute in Germany, carried out a "top down" calculation on how much wind energy there is, starting with the incoming solar radiation that drives the winds by creating temperature differences in the atmosphere. He concluded that somewhere between 18 TW and 68 TW could be extracted.

Cristina Archer and Mark Z. Jacobson presented a "bottom-up" estimate, which unlike Kleidon's are based on actual measurements of wind speeds, and found that there is 1700 TW of wind power at an altitude of 100 metres over land and sea. Of this, "between 72 and 170 TW could be extracted in a practical and cost-competitive manner". They later estimated 80 TW. However research at Harvard University estimates 1 Watt/m² on average and 2–10 MW/km² capacity for large scale wind farms, suggesting that these estimates of total global wind resources are too high by a factor of about 4.

The strength of wind varies, and an average value for a given location does not alone indicate the amount of energy a wind turbine could produce there.

To assess prospective wind power sites a probability distribution function is often fit to the observed wind speed data. Different locations will have different wind speed distributions. The

Weibull model closely mirrors the actual distribution of hourly/ten-minute wind speeds at many locations. The Weibull factor is often close to 2 and therefore a Rayleigh distribution can be used as a less accurate, but simpler model.

Wind Turbine

A wind turbine is a device that converts the wind's kinetic energy into electrical power.

Wind turbines are manufactured in a wide range of vertical and horizontal axis types. The smallest turbines are used for applications such as battery charging for auxiliary power for boats or caravans or to power traffic warning signs. Slightly larger turbines can be used for making contributions to a domestic power supply while selling unused power back to the utility supplier via the electrical grid. Arrays of large turbines, known as wind farms, are becoming an increasingly important source of renewable energy and are used by many countries as part of a strategy to reduce their reliance on fossil fuels.

Offshore wind farm, using 5 MW turbines REpower 5M in the North Sea off the coast of Belgium.

History

Windmills were used in Persia (present-day Iran) about 500-900 A.D. The windwheel of Hero of Alexandria marks one of the first known instances of wind powering a machine in history. However, the first known practical windmills were built in Sistan, an Eastern province of Iran, from the 7th century. These "Panemone" were vertical axle windmills, which had long vertical drive shafts with rectangular blades. Made of six to twelve sails covered in reed matting or cloth material, these windmills were used to grind grain or draw up water, and were used in the gristmilling and sugarcane industries.

Windmills first appeared in Europe during the Middle Ages. The first historical records of their use in England date to the 11th or 12th centuries and there are reports of German crusaders taking their windmill-making skills to Syria around 1190. By the 14th century, Dutch windmills were in use to drain areas of the Rhine delta. Advanced wind mills were described by Croatian inventor Fausto Veranzio. In his book Machinae Novae (1595) he described vertical axis wind turbines with curved or V-shaped blades.

James Blyth's electricity-generating wind turbine, photographed in 1891

The first electricity-generating wind turbine was a battery charging machine installed in July 1887 by Scottish academic James Blyth to light his holiday home in Marykirk, Scotland. Some months later American inventor Charles F. Brush was able to build the first automatically operated wind turbine after consulting local University professors and colleagues Jacob S. Gibbs and Brinsley Coleberd and successfully getting the blueprints peer-reviewed for electricity production in Cleveland, Ohio. Although Blyth's turbine was considered uneconomical in the United Kingdom electricity generation by wind turbines was more cost effective in countries with widely scattered populations.

In Denmark by 1900, there were about 2500 windmills for mechanical loads such as pumps and mills, producing an estimated combined peak power of about 30 MW. The largest machines were on 24-meter (79 ft) towers with four-bladed 23-meter (75 ft) diameter rotors. By 1908 there were 72 wind-driven electric generators operating in the United States from 5 kW to 25 kW. Around the time of World War I, American windmill makers were producing 100,000 farm windmills each year, mostly for water-pumping.

By the 1930s, wind generators for electricity were common on farms, mostly in the United States where distribution systems had not yet been installed. In this period, high-tensile steel was cheap, and the generators were placed atop prefabricated open steel lattice towers.

A forerunner of modern horizontal-axis wind generators was in service at Yalta, USSR in 1931. This was a 100 kW generator on a 30-meter (98 ft) tower, connected to the local 6.3 kV distribution system. It was reported to have an annual capacity factor of 32 percent, not much different from current wind machines.

In the autumn of 1941, the first megawatt-class wind turbine was synchronized to a utility grid in Vermont. The Smith-Putnam wind turbine only ran for 1,100 hours before suffering a critical failure. The unit was not repaired, because of shortage of materials during the war.

The first utility grid-connected wind turbine to operate in the UK was built by John Brown & Company in 1951 in the Orkney Islands.

Despite these diverse developments, developments in fossil fuel systems almost entirely eliminated any wind turbine systems larger than supermicro size. In the early 1970s, however, anti-nuclear protests in Denmark spurred artisan mechanics to develop microturbines of 22 kW. Organizing owners into associations and co-operatives lead to the lobbying of the government and utilities and provided incentives for larger turbines throughout the 1980s and later. Local activists in Germany, nascent turbine manufacturers in Spain, and large investors in the United States in the early 1990s then lobbied for policies that stimulated the industry in those countries. Later companies formed in India and China. As of 2012, Danish company Vestas is the world's biggest wind-turbine manufacturer.

Resources

A quantitative measure of the wind energy available at any location is called the Wind Power Density (WPD). It is a calculation of the mean annual power available per square meter of swept area of a turbine, and is tabulated for different heights above ground. Calculation of wind power density includes the effect of wind velocity and air density. Color-coded maps are prepared for a particular area described, for example, as "Mean Annual Power Density at 50 Metres". In the United States, the results of the above calculation are included in an index developed by the National Renewable Energy Laboratory and referred to as "NREL CLASS". The larger the WPD, the higher it is rated by class. Classes range from Class 1 (200 watts per square meter or less at 50 m altitude) to Class 7 (800 to 2000 watts per square m). Commercial wind farms generally are sited in Class 3 or higher areas, although isolated points in an otherwise Class 1 area may be practical to exploit.

Nordex N117/2400 in Germany, a modern low-wind turbine.

Wind turbines at the Jepirachí Eolian Park in La Guajira, Colombia.

Wind turbines are classified by the wind speed they are designed for, from class I to class IV, with A or B referring to the turbulence.

Class	Avg Wind Speed (m/s)	Turbulence
IA	10	18%
IB	10	16%
IIA	8.5	18%
IIB	8.5	16%
IIIA	7.5	18%
IIIB	7.5	16%
IVA	6	18%
IVB	6	16%

Efficiency

Not all the energy of blowing wind can be used, but some small wind turbines are designed to work at low wind speeds.

Conservation of mass requires that the amount of air entering and exiting a turbine must be equal. Accordingly, Betz's law gives the maximal achievable extraction of wind power by a wind turbine as 16/27 (59.3%) of the total kinetic energy of the air flowing through the turbine.

The maximum theoretical power output of a wind machine is thus 0.59 times the kinetic energy of the air passing through the effective disk area of the machine. If the effective area of the disk is A, and the wind velocity v, the maximum theoretical power output P is: $P = 0.59 \frac{1}{2} \rho v^3 A$

where ρ is air density

As wind is free (no fuel cost), wind-to-rotor efficiency (including rotor blade friction and drag) is one of many aspects impacting the final price of wind power. Further inefficiencies, such as gearbox losses, generator and converter losses, reduce the power delivered by a wind turbine. To protect components from undue wear, extracted power is held constant above the rated operating speed as theoretical power increases at the cube of wind speed, further reducing theoretical efficiency. In 2001, commercial utility-connected turbines deliver 75% to 80% of the Betz limit of power extractable from the wind, at rated operating speed.

Efficiency can decrease slightly over time due to wear. Analysis of 3128 wind turbines older than 10 years in Denmark showed that half of the turbines had no decrease, while the other half saw a production decrease of 1.2% per year.

Types

Wind turbines can rotate about either a horizontal or a vertical axis, the former being both older and more common. They can also include blades (transparent or not) or be bladeless.

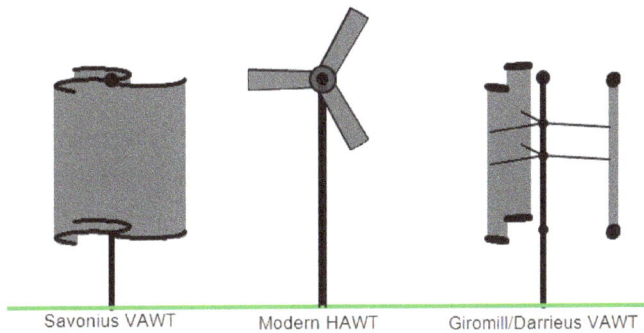

Savonius VAWT   Modern HAWT   Giromill/Darrieus VAWT

The three primary types: VAWT Savonius, HAWT towered; VAWT Darrieus as they appear in operation

Horizontal Axis

Horizontal-axis wind turbines (HAWT) have the main rotor shaft and electrical generator at the top of a tower, and must be pointed into the wind. Small turbines are pointed by a simple wind vane, while large turbines generally use a wind sensor coupled with a servo motor. Most have a gearbox, which turns the slow rotation of the blades into a quicker rotation that is more suitable to drive an electrical generator.

A turbine blade convoy passing through Edenfield, UK

Since a tower produces turbulence behind it, the turbine is usually positioned upwind of its supporting tower. Turbine blades are made stiff to prevent the blades from being pushed into the tower by high winds. Additionally, the blades are placed a considerable distance in front of the tower and are sometimes tilted forward into the wind a small amount.

Downwind machines have been built, despite the problem of turbulence (mast wake), because they don't need an additional mechanism for keeping them in line with the wind, and because in high winds the blades can be allowed to bend which reduces their swept area and thus their wind resistance. Since cyclical (that is repetitive) turbulence may lead to fatigue failures, most HAWTs are of upwind design.

Turbines used in wind farms for commercial production of electric power are usually three-bladed and pointed into the wind by computer-controlled motors. These have high tip speeds of over 320 km/h (200 mph), high efficiency, and low torque ripple, which contribute to good reliability. The blades are usually colored white for daytime visibility by aircraft and range in length from 20 to 40 meters (66 to 131 ft) or more. The tubular steel towers range from 60 to 90 meters (200 to 300 ft) tall.

The blades rotate at 10 to 22 revolutions per minute. At 22 rotations per minute the tip speed exceeds 90 meters per second (300 ft/s). A gear box is commonly used for stepping up the speed of the generator, although designs may also use direct drive of an annular generator. Some models operate at constant speed, but more energy can be collected by variable-speed turbines which use a solid-state power converter to interface to the transmission system. All turbines are equipped with protective features to avoid damage at high wind speeds, by feathering the blades into the wind which ceases their rotation, supplemented by brakes.

Year by year the size and height of turbines increase. Offshore wind turbines are built up to 8MW today and have a blade length up to 80m. Onshore wind turbines are installed in low wind speed areas and getting higher and higher towers. Usual towers of multi megawatt turbines have a height of 70 m to 120 m and in extremes up to 160 m, with blade tip speeds reaching 80 m/s to 90 m/s. Higher tip speeds means more noise and blade erosion.

Vertical Axis Design

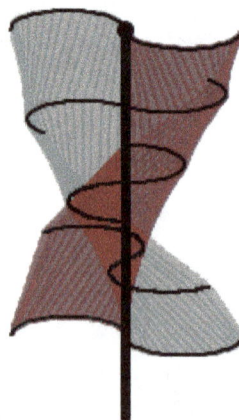

A vertical axis Twisted Savonius type turbine.

Vertical-axis wind turbines (or VAWTs) have the main rotor shaft arranged vertically. One advantage of this arrangement is that the turbine does not need to be pointed into the wind to be effective, which is an advantage on a site where the wind direction is highly variable. It is also an advantage when the turbine is integrated into a building because it is inherently less steerable. Also, the generator and gearbox can be placed near the ground, using a direct drive from the rotor assembly to the ground-based gearbox, improving accessibility for maintenance.

The key disadvantages include the relatively low rotational speed with the consequential higher torque and hence higher cost of the drive train, the inherently lower power coefficient, the 360-degree rotation of the aerofoil within the wind flow during each cycle and hence the highly dynamic loading on the blade, the pulsating torque generated by some rotor designs on the drive train, and the difficulty of modelling the wind flow accurately and hence the challenges of analysing and designing the rotor prior to fabricating a prototype.

When a turbine is mounted on a rooftop the building generally redirects wind over the roof and this can double the wind speed at the turbine. If the height of a rooftop mounted turbine tower is approximately 50% of the building height it is near the optimum for maximum wind energy and minimum wind turbulence. Wind speeds within the built environment are generally much lower than at exposed rural sites, noise may be a concern and an existing structure may not adequately resist the additional stress.

Subtypes of the vertical axis design include:

Offshore Horizontal Axis Wind Turbines (HAWTs) at Scroby Sands Wind Farm, UK

Onshore Horizontal Axis Wind Turbines in Zhangjiakou, China

Darrieus Wind Turbine

"Eggbeater" turbines, or Darrieus turbines, were named after the French inventor, Georges Darrieus. They have good efficiency, but produce large torque ripple and cyclical stress on the tower, which contributes to poor reliability. They also generally require some external power source, or an additional Savonius rotor to start turning, because the starting torque is very low. The torque ripple is reduced by using three or more blades which results in greater solidity of the rotor. Solidity is measured by blade area divided by the rotor area. Newer Darrieus type turbines are not held up by guy-wires but have an external superstructure connected to the top bearing.

Giromill

A subtype of Darrieus turbine with straight, as opposed to curved, blades. The cycloturbine variety has variable pitch to reduce the torque pulsation and is self-starting. The advantages of variable pitch are: high starting torque; a wide, relatively flat torque curve; a higher coefficient of performance; more efficient operation in turbulent winds; and a lower blade speed ratio which lowers blade bending stresses. Straight, V, or curved blades may be used.

Savonius Wind Turbine

These are drag-type devices with two (or more) scoops that are used in anemometers, *Flettner* vents (commonly seen on bus and van roofs), and in some high-reliability low-efficiency power turbines. They are always self-starting if there are at least three scoops.

Twisted Savonius

Twisted Savonius is a modified savonius, with long helical scoops to provide smooth torque. This is often used as a rooftop windturbine and has even been adapted for ships.

Another type of vertical axis is the Parallel turbine, which is similar to the crossflow fan or centrifugal fan. It uses the ground effect. Vertical axis turbines of this type have been tried for many years: a unit producing 10 kW was built by Israeli wind pioneer Bruce Brill in the 1980s.

Vortexis

The most recent advancement in Vertical Axis Wind Turbines has been the Vortexis VAWT, utilizing a pre-swirled augmented vertical axis wind turbine (PA-VAWT) designed for the purpose of developing a high efficiency VAWT concept that keeps the advantages of VAWT's compact size, lack of bias as to incoming wind direction, easy deployment and low radar cross section for use in mobile applications for the military, referred to in Special Operations as "Black Swan."

Design and Construction

Wind turbines are designed to exploit the wind energy that exists at a location. Aerodynamic modeling is used to determine the optimum tower height, control systems, number of blades and blade shape.

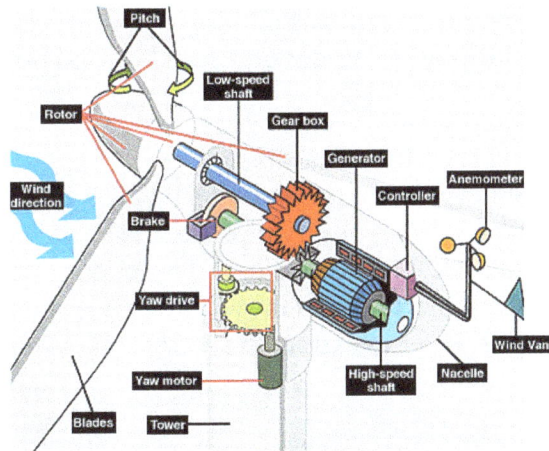

Components of a horizontal-axis wind turbine

Inside view of a wind turbine tower, showing the tendon cables.

Wind turbines convert wind energy to electricity for distribution. Conventional horizontal axis turbines can be divided into three components:

- The rotor component, which is approximately 20% of the wind turbine cost, includes the blades for converting wind energy to low speed rotational energy.

- The generator component, which is approximately 34% of the wind turbine cost, includes the electrical generator, the control electronics, and most likely a gearbox (e.g. planetary gearbox), adjustable-speed drive or continuously variable transmission component for converting the low speed incoming rotation to high speed rotation suitable for generating electricity.

- The structural support component, which is approximately 15% of the wind turbine cost, includes the tower and rotor yaw mechanism.

A 1.5 MW wind turbine of a type frequently seen in the United States has a tower 80 meters (260 ft) high. The rotor assembly (blades and hub) weighs 22,000 kilograms (48,000 lb). The nacelle, which contains the generator component, weighs 52,000 kilograms (115,000 lb). The concrete base for the tower is constructed using 26,000 kilograms (58,000 lb) of reinforcing steel and contains 190 cubic meters (250 cu yd) of concrete. The base is 15 meters (50 ft) in diameter and 2.4 meters (8 ft) thick near the center.

Among all renewable energy systems wind turbines have the highest effective intensity of power-harvesting surface because turbine blades not only harvest wind power, but also concentrate it.

Unconventional Designs

An E-66 wind turbine in the Windpark Holtriem, Germany, has an observation deck for visitors. Another turbine of the same type with an observation deck is located in Swaffham, England. Airborne wind turbine designs have been proposed and developed for many years but have yet to produce significant amounts of energy. In principle, wind turbines may also be used in conjunction with a large vertical solar updraft tower to extract the energy due to air heated by the sun.

The corkscrew shaped wind turbine at Progressive Field in Cleveland, Ohio

Wind turbines which utilise the Magnus effect have been developed.

A ram air turbine (RAT) is a special kind of small turbine that is fitted to some aircraft. When deployed, the RAT is spun by the airstream going past the aircraft and can provide power for the most essential systems if there is a loss of all on-board electrical power, as in the case of the "Gimli Glider".

The two-bladed turbine SCD 6MW offshore turbine designed by aerodyn Energiesysteme, built by MingYang Wind Power has a helideck for helicopters on top of its nacelle. The prototype was erected in 2014 in Rudong China.

Turbine Monitoring and Diagnostics

Due to data transmission problems, structural health monitoring of wind turbines is usually performed using several accelerometers and strain gages attached to the nacelle to monitor the gearbox and equipments. Currently, digital image correlation and stereophotogrammetry are used to measure dynamics of wind turbine blades. These methods usually measure displacement and strain to identify location of defects. Dynamic characteristics of non-rotating wind turbines have been measured using digital image correlation and photogrammetry. Three dimensional point tracking has also been used to measure rotating dynamics of wind turbines.

Materials and Durability

Currently serving wind turbine blades are mainly made of composite materials. These blades are usually made of a polyester resin, a vinyl resin, and epoxy thermosetting matrix resin and E-glass fibers,

S-glass fibers and carbon fiber reinforced materials. Construction may use manual layup techniques or composite resin injection molding. As the price of glass fibers is only about one tenth the price of carbon fiber, glass fiber is still dominant. One of the predominant ways wind turbines have gain performance is by increasing rotor diameters, and thus blade length. Longer blades place more demands on the strength and stiffness of the materials. Stiffness is especially important to avoid having blades flex to the degree that they hit the tower of the wind turbine. Carbon fiber is between 4 and 6 times stiffer than glass fiber, so carbon fiber is becoming more common in wind turbine blades.

Wind Turbines on Public Display

A few localities have exploited the attention-getting nature of wind turbines by placing them on public display, either with visitor centers around their bases, or with viewing areas farther away. The wind turbines are generally of conventional horizontal-axis, three-bladed design, and generate power to feed electrical grids, but they also serve the unconventional roles of technology demonstration, public relations, and education.

The Nordex N50 wind turbine and visitor centre of Lamma Winds in Hong Kong, China

Small Wind Turbines

Small wind turbines may be used for a variety of applications including on- or off-grid residences, telecom towers, offshore platforms, rural schools and clinics, remote monitoring and other purposes that require energy where there is no electric grid, or where the grid is unstable. Small wind turbines may be as small as a fifty-watt generator for boat or caravan use. Hybrid solar and wind powered units are increasingly being used for traffic signage, particularly in rural locations, as they avoid the need to lay long cables from the nearest mains connection point. The U.S. Department of Energy's National Renewable Energy Laboratory (NREL) defines small wind turbines as those smaller than or equal to 100 kilowatts. Small units often have direct drive generators, direct current output, aeroelastic blades, lifetime bearings and use a vane to point into the wind.

Larger, more costly turbines generally have geared power trains, alternating current output, flaps and are actively pointed into the wind. Direct drive generators and aeroelastic blades for large wind turbines are being researched.

Wind Turbine Spacing

On most horizontal windturbine farms, a spacing of about 6-10 times the rotor diameter is often upheld. However, for large wind farms distances of about 15 rotor diameters should be more economically optimal, taking into account typical wind turbine and land costs. This conclusion has been reached by research conducted by Charles Meneveau of the Johns Hopkins University, and Johan Meyers of Leuven University in Belgium, based on computer simulations that take into account the detailed interactions among wind turbines (wakes) as well as with the entire turbulent atmospheric boundary layer. Moreover, recent research by John Dabiri of Caltech suggests that vertical wind turbines may be placed much more closely together so long as an alternating pattern of rotation is created allowing blades of neighbouring turbines to move in the same direction as they approach one another.

Operability

Maintenance

Wind turbines need regular maintenance to stay reliable and available, reaching 98%.

Modern turbines usually have a small onboard crane for hoisting maintenance tools and minor components. However, large heavy components like generator, gearbox, blades and so on are rarely replaced and a heavy lift external crane is needed in those cases. If the turbine has a difficult access road, a containerized crane can be lifted up by the internal crane to provide heavier lifting.

Repowering

Installation of new wind turbines can be controversial. An alternative is repowering, where existing wind turbines are replaced with bigger, more powerful ones, sometimes in smaller numbers while keeping or increasing capacity.

Demolition

Older turbines were in some early cases not required to be removed when reaching the end of their life. Some still stand, waiting to be recycled or repowered.

A demolition industry develops to recycle offshore turbines at a cost of DKK 2–4 million per MW, to be guaranteed by the owner.

Records

Largest capacity conventional drive

> The Vestas V164 has a rated capacity of 8 MW, has an overall height of 220 m (722 ft), a diameter of 164 m (538 ft), is for offshore use, and is the world's largest-capacity wind turbine since its introduction in 2014. The conventional drive train consist of a main gearbox and a medium speed PM generator. Prototype installed in 2014 at the National Test Center Denmark nearby Østerild. Series production starts end of 2015.

Fuhrländer Wind Turbine Laasow, in Brandenburg, Germany, among the world's tallest wind turbines

Largest capacity direct drive

The Enercon E-126 with 7.58 MW and 127 m rotor diameter is the largest direct drive turbine. It's only for onshore use. The turbine has parted rotor blades with 2 sections for transport. In July 2016, Siemens upgraded its 7 to 8 MW.

Éole, the largest vertical axis wind turbine, in Cap-Chat, Quebec, Canada

Largest vertical-axis

Le Nordais wind farm in Cap-Chat, Quebec has a vertical axis wind turbine (VAWT) named Éole, which is the world's largest at 110 m. It has a nameplate capacity of 3.8 MW.

Largest 1-bladed turbine

Riva Calzoni M33 was a single-bladed wind turbine with 350 kW, designed and built In Bologna in 1993.

Largest 2-bladed turbine

Today's biggest 2-bladed turbine is build by Mingyang Wind Power in 2013. It is a SCD6.5MW offshore downwind turbine, designed by aerodyn Energiesysteme.

Largest swept area

The turbine with the largest swept area is the Samsung S7.0-171, with a diameter of 171 m, giving a total sweep of 22966 m².

Tallest

A Nordex 3.3 MW was installed in July 2016. It has a total height of 230m, and a hub height of 164m on 100m concrete tower bottom with steel tubes on top (hybrid tower).

Vestas V164 was the tallest wind turbine, standing in Østerild, Denmark, 220 meters tall, constructed in 2014. It has a steel tube tower.

Highest tower

Fuhrländer installed a 2.5MW turbine on a 160m lattice tower in 2003 (see Fuhrländer Wind Turbine Laasow and Nowy Tomyśl Wind Turbines).

Most rotors

Lagerwey has build Four-in-One, a multi rotor wind turbine with one tower and four rotors near Maasvlakte. In April 2016, Vestas installed a 900 kW quadrotor test wind turbine at Risø, made from 4 recycled 225 kW V29 turbines.

Most productive

Four turbines at Rønland wind farm in Denmark share the record for the most productive wind turbines, with each having generated 63.2 GWh by June 2010.

Highest-situated

Since 2013 the world's highest-situated wind turbine was made and installed by WindAid and is located at the base of the Pastoruri Glacier in Peru at 4,877 meters (16,001 ft) above sea level. The site uses the WindAid 2.5 kW wind generator to supply power to a small rural community of micro entrepreneurs who cater to the tourists who come to the Pastoruri glacier.

Largest floating wind turbine

The world's largest—and also the first operational deep-water *large-capacity*—floating wind turbine is the 2.3 MW Hywind currently operating 10 kilometers (6.2 mi) offshore in 220-meter-deep water, southwest of Karmøy, Norway. The turbine began operating in September 2009 and utilizes a Siemens 2.3 MW turbine.

Wind Turbine Design

Wind turbine design is the process of defining the form and specifications of a wind turbine to extract energy from the wind. A wind turbine installation consists of the necessary systems needed to capture the wind's energy, point the turbine into the wind, convert mechanical rotation into electrical power, and other systems to start, stop, and control the turbine.

An example of a wind turbine, this 3 bladed turbine is the classic design of modern wind turbines

This article covers the design of horizontal axis wind turbines (HAWT) since the majority of commercial turbines use this design.

In 1919 the physicist Albert Betz showed that for a hypothetical ideal wind-energy extraction machine, the fundamental laws of conservation of mass and energy allowed no more than 16/27 (59.3%) of the kinetic energy of the wind to be captured. This Betz' law limit can be approached by modern turbine designs which may reach 70 to 80% of this theoretical limit.

In addition to aerodynamic design of the blades, design of a complete wind power system must also address design of the hub, controls, generator, supporting structure and foundation. Further design questions arise when integrating wind turbines into electrical power grids.

Aerodynamics

The shape and dimensions of the blades of the wind turbine are determined by the aerodynamic performance required to efficiently extract energy from the wind, and by the strength required to resist the forces on the blade.

The aerodynamics of a horizontal-axis wind turbine are not straightforward. The air flow at the blades is not the same as the airflow far away from the turbine. The very nature of the way in which energy is extracted from the air also causes air to be deflected by the turbine. In addition the aerodynamics of a wind turbine at the rotor surface exhibit phenomena that are rarely seen in other aerodynamic fields.

In 1919 the physicist Albert Betz showed that for a hypothetical ideal wind-energy extraction machine, the fundamental laws of conservation of mass and energy allowed no more than 16/27 (59.3%) of the kinetic energy of the wind to be captured. This Betz' law limit can be approached by modern turbine designs which may reach 70 to 80% of this theoretical limit.

Wind rotor profile

Power Control

The speed at which a wind turbine rotates must be controlled for efficient power generation and to keep the turbine components within designed speed and torque limits. The centrifugal force on the spinning blades increases as the square of the rotation speed, which makes this structure sensitive to overspeed. Because the power of the wind increases as the cube of the wind speed, turbines have to be built to survive much higher wind loads (such as gusts of wind) than those from which they can practically generate power. Wind turbines have ways of reducing torque in high winds.

A wind turbine is designed to produce power over a range of wind speeds. All wind turbines are designed for a maximum wind speed, called the survival speed, above which they will be damaged. The survival speed of commercial wind turbines is in the range of 40 m/s (144 km/h, 89 MPH) to 72 m/s (259 km/h, 161 MPH). The most common survival speed is 60 m/s (216 km/h, 134 MPH).

If the rated wind speed is exceeded the power has to be limited. There are various ways to achieve this.

A control system involves three basic elements: sensors to measure process variables, actuators to manipulate energy capture and component loading, and control algorithms to coordinate the actuators based on information gathered by the sensors.

Stall

Stalling works by increasing the angle at which the relative wind strikes the blades (angle of attack), and it reduces the induced drag (drag associated with lift). Stalling is simple because it can be made to happen passively (it increases automatically when the winds speed up), but it increases the cross-section of the blade face-on to the wind, and thus the ordinary drag. A fully stalled turbine blade, when stopped, has the flat side of the blade facing directly into the wind.

A fixed-speed HAWT (Horizontal Axis Wind Turbine) inherently increases its angle of attack at higher wind speed as the blades speed up. A natural strategy, then, is to allow the blade to stall when the wind speed increases. This technique was successfully used on many early HAWTs. However, on some of these blade sets, it was observed that the degree of blade pitch tended to increase audible noise levels.

Vortex generators may be used to control the lift characteristics of the blade. The VGs are placed on the airfoil to enhance the lift if they are placed on the lower (flatter) surface or limit the maximum lift if placed on the upper (higher camber) surface.

Furling works by decreasing the angle of attack, which reduces the induced drag from the lift of the rotor, as well as the cross-section. One major problem in designing wind turbines is getting the blades to stall or furl quickly enough should a gust of wind cause sudden acceleration. A fully furled turbine blade, when stopped, has the edge of the blade facing into the wind.

Loads can be reduced by making a structural system softer or more flexible. This could be accomplished with downwind rotors or with curved blades that twist naturally to reduce angle of attack at higher wind speeds. These systems will be nonlinear and will couple the structure to the flow field - thus, design tools must evolve to model these nonlinearities.

Standard modern turbines all furl the blades in high winds. Since furling requires acting against the torque on the blade, it requires some form of pitch angle control, which is achieved with a slewing drive. This drive precisely angles the blade while withstanding high torque loads. In addition, many turbines use hydraulic systems. These systems are usually spring-loaded, so that if hydraulic power fails, the blades automatically furl. Other turbines use an electric servomotor for every rotor blade. They have a small battery-reserve in case of an electric-grid breakdown. Small wind turbines (under 50 kW) with variable-pitching generally use systems operated by centrifugal force, either by flyweights or geometric design, and employ no electric or hydraulic controls.

Fundamental gaps exist in pitch control, limiting the reduction of energy costs, according to a report from a coalition of researchers from universities, industry, and government, supported by the Atkinson Center for a Sustainable Future. Load reduction is currently focused on full-span blade pitch control, since individual pitch motors are the actuators currently available on commercial turbines. Significant load mitigation has been demonstrated in simulations for blades, tower, and drive train. However, there is still research needed, the methods for realization of full-span blade pitch control need to be developed in order to increase energy capture and mitigate fatigue loads.

A control technique applied to the pitch angle is done by comparing the current active power of the engine with the value of active power at the rated engine speed (active power reference, Ps reference). Control of the pitch angle in this case is done with a PI controller controls. However, in order to have a realistic response to the control system of the pitch angle, the actuator uses the time constant Tservo, an integrator and limiters so as the pitch angle to be from 0° to 30° with a rate of change (± 10° per sec).

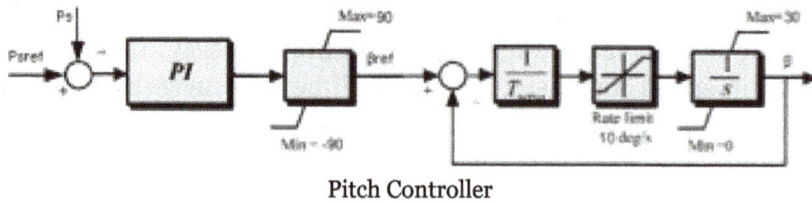

Pitch Controller

From the figure at the right, the reference pitch angle is compared with the actual pitch angle b and then the error is corrected by the actuator. The reference pitch angle, which comes from the PI controller, goes through a limiter. Restrictions on limits are very important to maintain the pitch angle in real term. Limiting the rate of change is very important especially during faults in the network. The importance is due to the fact that the controller decides how quickly it can reduce the aerodynamic energy to avoid acceleration during errors.

Other Controls

Generator Torque

Modern large wind turbines are variable-speed machines. When the wind speed is below rated, generator torque is used to control the rotor speed in order to capture as much power as possible. The most power is captured when the tip speed ratio is held constant at its optimum value (typically 6 or 7). This means that as wind speed increases, rotor speed should increase proportionally. The difference between the aerodynamic torque captured by the blades and the applied generator torque controls the rotor speed. If the generator torque is lower, the rotor accelerates, and if the generator torque is higher, the rotor slows down. Below rated wind speed, the generator torque control is active while the blade pitch is typically held at the constant angle that captures the most power, fairly flat to the wind. Above rated wind speed, the generator torque is typically held constant while the blade pitch is active.

One technique to control a permanent magnet synchronous motor is Field Oriented Control. Field Oriented Control is a closed loop strategy composed of two current controllers (an inner loop and outer loop cascade design) necessary for controlling the torque, and one speed controller.

Constant torque angle control

In this control strategy the d axis current is kept zero, while the vector current is align with the q axis in order to maintain the torque angle equal with 90°. This is one of the most used control strategy because of the simplicity, by controlling only the Iqs current. So, now the electromagnetic torque equation of the permanent magnet synchronous generator is simply a linear equation depend on the Iqs current only.

So, the electromagnetic torque for Ids = 0 (we can achieve that with the d-axis controller) is now:

$$T_e = 3/2\ p\ (\lambda_{pm}\ I_{qs} + (L_{ds} - L_{qs})\ I_{ds}\ I_{qs}) = 3/2\ p\ \lambda_{pm}\ I_{qs}$$

So, the complete system of the machine side converter and the cascaded PI controller loops is given by the figure in the right. In that we have the control inputs, which are the duty rations m_{ds} and m_{qs}, of the PWM-regulated converter. Also, we can see the control scheme for the wind turbine in the machine side and simultaneously how we keep the I_{ds} zero (the electromagnetic torque equation is linear).

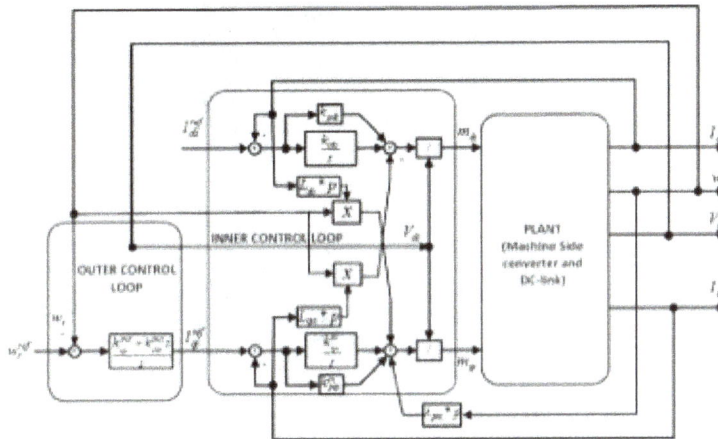

Machine Side Controller Design

Yawing

Modern large wind turbines are typically actively controlled to face the wind direction measured by a wind vane situated on the back of the nacelle. By minimizing the yaw angle (the misalignment between wind and turbine pointing direction), the power output is maximized and non-symmetrical loads minimized. However, since the wind direction varies quickly the turbine will not strictly follow the direction and will have a small yaw angle on average. The power output losses can simply be approximated to fall with (cos(yaw angle))³. Particularly at low-to-medium wind speeds, yawing can make a significant reduction in turbine output, with wind direction variations of ±30° being quite common and long response times of the turbines to changes in wind direction. At high wind speeds, the wind direction is less variable.

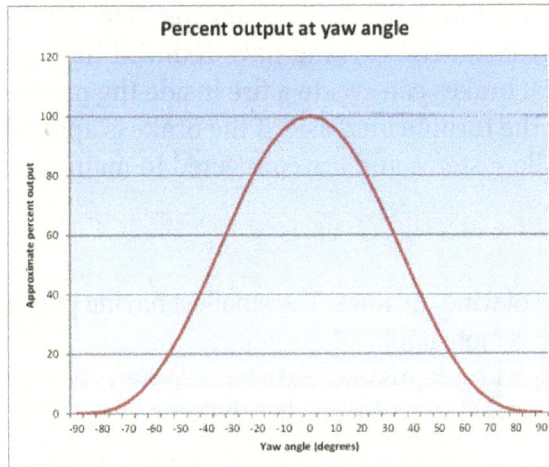

Percent output vs. wind angle

Electrical Braking

Braking of a small wind turbine can be done by dumping energy from the generator into a resistor bank, converting the kinetic energy of the turbine rotation into heat. This method is useful if the kinetic load on the generator is suddenly reduced or is too small to keep the turbine speed within its allowed limit.

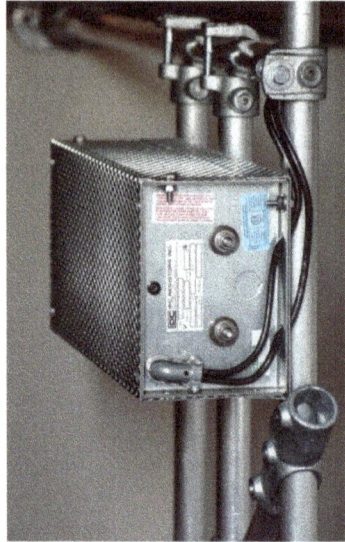

2kW Dynamic braking resistor for small wind turbine.

Cyclically braking causes the blades to slow down, which increases the stalling effect, reducing the efficiency of the blades. This way, the turbine's rotation can be kept at a safe speed in faster winds while maintaining (nominal) power output. This method is usually not applied on large grid-connected wind turbines.

Mechanical Braking

A mechanical drum brake or disk brake is used to stop turbine in emergency situation such as extreme gust events or over speed. This brake is a secondary means to hold the turbine at rest for maintenance, with a rotor lock system as primary means. Such brakes are usually applied only after blade furling and electromagnetic braking have reduced the turbine speed generally 1 or 2 rotor RPM, as the mechanical brakes can create a fire inside the nacelle if used to stop the turbine from full speed. The load on the turbine increases if the brake is applied at rated RPM. Mechanical brakes are driven by hydraulic systems and are connected to main control box.

Turbine Size

There are different size classes of wind turbines. The smallest having power production less than 10 kW are used in homes, farms and remote applications whereas intermediate wind turbines (10-250 kW) are useful for village power, hybrid systems and distributed power. The largest wind turbines (660 kW – 2+MW) are used in central station wind farms, distributed power and community wind.

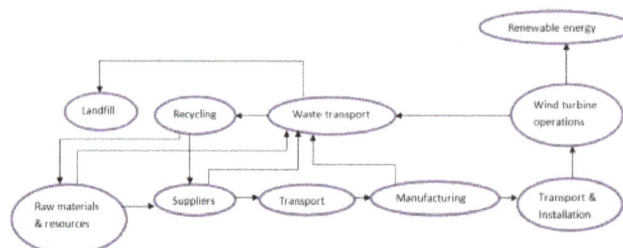

Flow diagram for wind turbine plant

For a given survivable wind speed, the mass of a turbine is approximately proportional to the cube of its blade-length. Wind power intercepted by the turbine is proportional to the square of its blade-length. The maximum blade-length of a turbine is limited by both the strength and stiffness of its material.

A person standing beside 15 m long blades.

Labor and maintenance costs increase only gradually with increasing turbine size, so to minimize costs, wind farm turbines are basically limited by the strength of materials, and siting requirements.

Typical modern wind turbines have diameters of 40 to 90 metres (130 to 300 ft) and are rated between 500 kW and 2 MW. As of 2014 the most powerful turbine, the Vestas V-164, is rated at 8 MW and has a rotor diameter of 164m.

Nacelle

The nacelle is housing the gearbox and generator connecting the tower and rotor. Sensors detect the wind speed and direction, and motors turn the nacelle into the wind to maximize output.

Gearbox

In conventional wind turbines, the blades spin a shaft that is connected through a gearbox to the generator. The gearbox converts the turning speed of the blades 15 to 20 rotations per minute for a large, one-megawatt turbine into the faster 1,800 rotations per minute that the generator needs to generate electricity. Analysts from GlobalData estimate that gearbox market grows from $3.2bn in 2006 to $6.9bn in 2011, and to $8.1bn by 2020. Market leaders were Winergy in 2011. The use of magnetic gearboxes has also been explored as a way of reducing wind turbine maintenance costs.

Generator

For large, commercial size horizontal-axis wind turbines, the electrical generator is mounted in a nacelle at the top of a tower, behind the hub of the turbine rotor. Typically wind turbines generate electricity through asynchronous machines that are directly connected with the electricity grid. Usually the rotational speed of the wind turbine is slower than the equivalent rotation speed of the electrical network: typical rotation speeds for wind generators are 5–20 rpm while a directly connected machine will have an electrical speed between 750 and 3600 rpm. Therefore, a gearbox is inserted between the rotor hub and the generator. This also reduces the generator cost and weight. Commercial size generators have a rotor carrying a field winding so that a rotating magnetic field is produced inside a set of windings called the stator. While the rotating field winding consumes a fraction of a percent of the generator output, adjustment of the field current allows good control over the generator output voltage.

Older style wind generators rotate at a constant speed, to match power line frequency, which allowed the use of less costly induction generators. Newer wind turbines often turn at whatever speed generates electricity most efficiently. The varying output frequency and voltage can be matched to the fixed values of the grid using multiple technologies such as doubly fed induction generators or full-effect converters where the variable frequency current produced is converted to DC and then back to AC. Although such alternatives require costly equipment and cause power loss, the turbine can capture a significantly larger fraction of the wind energy. In some cases, especially when turbines are sited offshore, the DC energy will be transmitted from the turbine to a central (onshore) inverter for connection to the grid.

Gearless Wind Turbine

Gearless wind turbines (also called direct drive) get rid of the gearbox completely. Instead, the rotor shaft is attached directly to the generator, which spins at the same speed as the blades. Enercon and EWT (Formerly known as Lagerwey) have produced gearless wind turbines with separately electrically excited generators for many years, and Siemens produces a gearless "inverted generator" 3 MW model while developing a 6 MW model. To make up for a direct drive generator's slower spinning rate, the diameter of the generator's rotor is increased hence containing more magnets which lets it create a lot of power when turning slowly.

Gearless wind turbines are often heavier than gear based wind turbines. A study by the EU called "Reliawind" based on the largest sample size of turbines, has shown that the reliability of gearboxes is not the main problem in wind turbines. The reliability of direct drive turbines offshore is still not known, since the sample size is so small.

Experts from Technical University of Denmark estimate that a geared generator with permanent magnets may use 25 kg/MW of the rare earth element Neodymium, while a gearless may use 250 kg/MW.

In December 2011, the US Department of Energy published a report stating critical shortage of rare earth elements such as neodymium used in large quantities for permanent magnets in gearless wind turbines. China produces more than 95% of rare earth elements, while Hitachi holds more than 600 patents covering Neodymium magnets. Direct-drive turbines require 600 kg of permanent magnet material per megawatt, which translates to several hundred kilograms of rare earth content per megawatt, as neodymium content is estimated to be 31% of magnet weight. Hybrid drivetrains (intermediate between direct drive and traditional geared) use significantly less rare earth materials. While permanent magnet wind turbines only account for about 5% of the market outside of China, their market share inside of China is estimated at 25% or higher. In 2011, demand for neodymium in wind turbines was estimated to be 1/5 of that in electric vehicles.

Blades

Blade Design

The ratio between the speed of the blade tips and the speed of the wind is called tip speed ratio. High efficiency 3-blade-turbines have tip speed/wind speed ratios of 6 to 7. Modern wind turbines are designed to spin at varying speeds (a consequence of their generator design). Use of aluminum and composite materials in their blades has contributed to low rotational inertia, which means that newer wind turbines can accelerate quickly if the winds pick up, keeping the tip speed ratio more nearly constant. Operating closer to their optimal tip speed ratio during energetic gusts of wind allows wind turbines to improve energy capture from sudden gusts that are typical in urban settings.

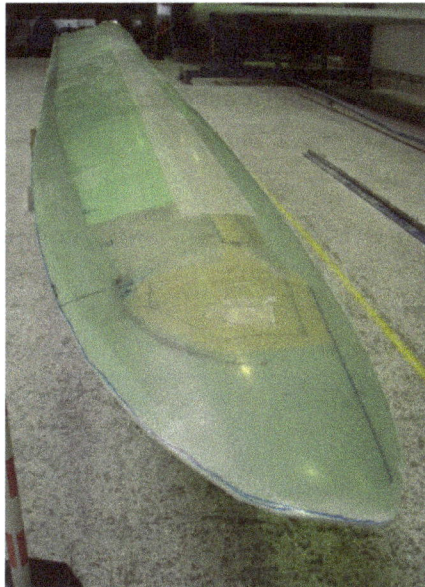

Unpainted tip of a blade

In contrast, older style wind turbines were designed with heavier steel blades, which have higher inertia, and rotated at speeds governed by the AC frequency of the power lines. The high inertia buffered the changes in rotation speed and thus made power output more stable.

It is generally understood that noise increases with higher blade tip speeds. To increase tip speed without increasing noise would allow reduction the torque into the gearbox and generator and reduce overall structural loads, thereby reducing cost. The reduction of noise is linked to the detailed aerodynamics of the blades, especially factors that reduce abrupt stalling. The inability to predict stall restricts the development of aggressive aerodynamic concepts. Some blades (mostly on Enercon) have a winglet to increase performance and/or reduce noise.

A blade can have a lift-to-drag ratio of 120, compared to 70 for a sailplane and 15 for an airliner.

The Hub

In simple designs, the blades are directly bolted to the hub and are unable to pitch, which leads to aerodynamic stall above certain windspeeds. In other more sophisticated designs, they are bolted to the pitch mechanism, which adjusts their angle of attack according to the wind speed to control their rotational speed. The pitch mechanism is itself bolted to the hub. The hub is fixed to the rotor shaft which drives the generator directly or through a gearbox.

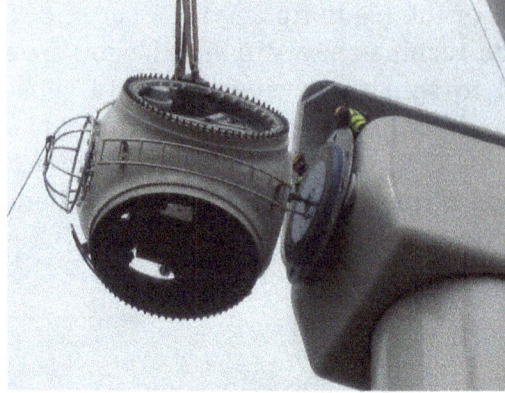

A Wind turbine hub being installed

Blade Count

The number of blades is selected for aerodynamic efficiency, component costs, and system reliability. Noise emissions are affected by the location of the blades upwind or downwind of the tower and the speed of the rotor. Given that the noise emissions from the blades' trailing edges and tips vary by the 5th power of blade speed, a small increase in tip speed can make a large difference.

The 98 meter diameter, two-bladed NASA/DOE Mod-5B wind turbine was the largest operating wind turbine in the world in the early 1990s

Wind turbines developed over the last 50 years have almost universally used either two or three blades. However, there are patents that present designs with additional blades, such as Chan Shin's Multi-unit rotor blade system integrated wind turbine. Aerodynamic efficiency increases with number of blades but with diminishing return. Increasing the number of blades from one to two yields a six percent increase in aerodynamic efficiency, whereas increasing the blade count from two to three yields only an additional three percent in efficiency. Further increasing the blade count yields minimal improvements in aerodynamic efficiency and sacrifices too much in blade stiffness as the blades become thinner.

Theoretically, an infinite number of blades of zero width is the most efficient, operating at a high value of the tip speed ratio. But other considerations lead to a compromise of only a few blades.

The NASA test of a one-bladed wind turbine rotor configuration at Plum Brook Station near Sandusky, Ohio

Component costs that are affected by blade count are primarily for materials and manufacturing of the turbine rotor and drive train. Generally, the lower the number of blades, the lower the material and manufacturing costs will be. In addition, the lower the number of blades, the higher the rotational speed can be. This is because blade stiffness requirements to avoid interference with the tower limit how thin the blades can be manufactured, but only for upwind machines; deflection of blades in a downwind machine results in increased tower clearance. Fewer blades with higher rotational speeds reduce peak torques in the drive train, resulting in lower gearbox and generator costs.

System reliability is affected by blade count primarily through the dynamic loading of the rotor into the drive train and tower systems. While aligning the wind turbine to changes in wind direction (yawing), each blade experiences a cyclic load at its root end depending on blade position. This is true of one, two, three blades or more. However, these cyclic loads when combined together at the drive train shaft are symmetrically balanced for three blades, yielding smoother operation during turbine yaw. Turbines with one or two blades can use a pivoting teetered hub to also nearly eliminate the cyclic loads into the drive shaft and system during yawing. A Chinese 3.6 MW two-blade is being tested in Denmark. Mingyang won a bid for 87 MW (29 * 3 MW) two-bladed offshore wind turbines near Zhuhai in 2013.

Finally, aesthetics can be considered a factor in that some people find that the three-bladed rotor is more pleasing to look at than a one- or two-bladed rotor.

Blade Materials

Wood and canvas sails were used on early windmills due to their low price, availability, and ease of manufacture. Smaller blades can be made from light metals such as aluminium. These materials, however, require frequent maintenance. Wood and canvas construction limits the airfoil shape to a flat plate, which has a relatively high ratio of drag to force captured (low aerodynamic efficiency) compared to solid airfoils. Construction of solid airfoil designs requires inflexible materials such as metals or composites. Some blades also have incorporated lightning conductors.

Several modern wind turbines use rotor blades with carbon-fibre girders to reduce weight.

In general, ideal materials should meet the following criteria:

- wide availability and easy processing to reduce cost and maintenance
- low weight or density to reduce gravitational forces
- high strength to withstand strong loading of wind and gravitational force of the blade itself
- high fatigue resistance to withstand cyclic loading
- high stiffness to ensure stability of the optimal shape and orientation of the blade and clearance with the tower
- high fracture toughness
- the ability to withstand environmental impacts such as lightning strikes, humidity, and temperature

New wind turbine designs push power generation from the single megawatt range to upwards of 10 megawatts using larger and larger blades. A larger area effectively increases the tip-speed ratio of a turbine at a given wind speed, thus increasing its energy extraction. Computer-aided engineering software such as HyperSizer (originally developed for spacecraft design) can be used to improve blade design.

As of 2015 the rotor diameters of onshore wind turbine blades are as large as 130 meters, while the diameter of offshore turbines reach 170 meters. In 2001, an estimated 50 million kilograms of fibreglass laminate were used in wind turbine blades.

An important goal of larger blade systems is to control blade weight. Since blade mass scales as the cube of the turbine radius, loading due to gravity constrains systems with larger blades. Gravitational loads include axial and tensile/ compressive loads (top/bottom of rotation) as well as bending (lateral positions). The magnitude of these loads fluctuates cyclically and the edge-wise moments are reversed every 180° of rotation. Typical rotor speeds and design life are ~10rpm and 20 years, respectively, with the number of lifetime revolutions on the order of 10^8. Considering wind, it is expected that turbine blades go through ~10^9 loading cycles. Wind is another source of rotor blade loading. Lift causes bending in the flapwise direction (out of rotor plane) while air flow around the blade cause edgewise bending (in the rotor plane). Flapwise bending involves tension on the pressure (upwind) side and compression on the suction (downwind) side. Edgewise bending involves tension on the leading edge and compression on the trailing edge.

Wind loads are cyclical because of natural variability in wind speed and wind shear (higher speeds at top of rotation).

Failure in ultimate loading of wind-turbine rotor blades exposed to wind and gravity loading is a failure mode that needs to be considered when the rotor blades are designed. The wind speed that causes bending of the rotor blades exhibits a natural variability, and so does the stress response in the rotor blades. Also, the resistance of the rotor blades, in terms of their tensile strengths, exhibits a natural variability.

In light of these failure modes and increasingly larger blade systems, there has been continuous effort toward developing cost-effective materials with higher strength-to-mass ratios. In order to extend the current 20 year lifetime of blades and enable larger area blades to be cost-effective, the design and materials need to be optimized for stiffness, strength, and fatigue resistance.

The majority of current commercialized wind turbine blades are made from fiber-reinforced polymers (FRP's), which are composites consisting of a polymer matrix and fibers. The long fibers provide longitudinal stiffness and strength, and the matrix provides fracture toughness, delamination strength, out-of-plane strength, and stiffness. Material indices based on maximizing power efficiency, and having high fracture toughness, fatigue resistance, and thermal stability, have been shown to be highest for glass and carbon fiber reinforced plastics (GFRP's and CFRPs).

Fiberglass-reinforced epoxy blades of Siemens SWT-2.3-101 wind turbines. The blade size of 49 meters is in comparison to a substation behind them at Wolfe Island Wind Farm.

Manufacturing blades in the 40 to 50 metre range involves proven fibreglass composite fabrication techniques. Manufactures such as Nordex SE and GE Wind use an infusion process. Other manufacturers use variations on this technique, some including carbon and wood with fibreglass in an epoxy matrix. Other options include preimpregnated ("prepreg") fibreglass and vacuum-assisted resin transfer molding. Each of these options use a glass-fibre reinforced polymer composite constructed with differing complexity. Perhaps the largest issue with more simplistic, open-mould, wet systems are the emissions associated with the volatile organics released. Preimpregnated materials and resin infusion techniques avoid the release of volatiles by containing all VOC's. However, these contained processes have their own challenges, namely the production of thick laminates necessary for structural components becomes more difficult. As the preform resin permeability dictates the maximum laminate thickness, bleeding is required to eliminate voids and ensure proper resin distribution. One solution to resin distribution a partially preimpregnated fibreglass.

During evacuation, the dry fabric provides a path for airflow and, once heat and pressure are applied, resin may flow into the dry region resulting in a thoroughly impregnated laminate structure.

Epoxy-based composites have environmental, production, and cost advantages over other resin systems. Epoxies also allow shorter cure cycles, increased durability, and improved surface finish. Prepreg operations further reduce processing time over wet lay-up systems. As turbine blades pass 60 metres, infusion techniques become more prevalent; the traditional resin transfer moulding injection time is too long as compared to the resin set-up time, limiting laminate thickness. Injection forces resin through a thicker ply stack, thus depositing the resin where in the laminate structure before gelation occurs. Specialized epoxy resins have been developed to customize lifetimes and viscosity.

Carbon fibre-reinforced load-bearing spars can reduce weight and increase stiffness. Using carbon fibres in 60 metre turbine blades is estimated to reduce total blade mass by 38% and decrease cost by 14% compared to 100% fibreglass. Carbon fibres have the added benefit of reducing the thickness of fiberglass laminate sections, further addressing the problems associated with resin wetting of thick lay-up sections. Wind turbines may also benefit from the general trend of increasing use and decreasing cost of carbon fibre materials.

Although glass and carbon fibers have many optimal qualities for turbine blade performance, there are several downsides to these current fillers, including the fact that high filler fraction (10-70 wt%) causes increased density as well as microscopic defects and voids that often lead to premature failure.

Recent developments include interest in using carbon nanotubes (CNT's) to reinforce polymer-based nanocomposites. CNT's can be grown or deposited on the fibers, or added into polymer resins as a matrix for FRP structures. Using nanoscale CNT's as filler instead of traditional microscale filler (such as glass or carbon fibers) results in CNT/polymer nanocomposites, for which the properties can be changed significantly at very low filler contents (typically < 5 wt%). They have very low density, and improve the elastic modulus, strength, and fracture toughness of the polymer matrix. The addition of CNT's to the matrix also reduces the propagation of interlaminar cracks which can be a problem in traditional FRP's.

Further improvement is possible through the use of carbon nanofibers (CNF's) in the blade coatings. A major problem in desert environments is erosion of the leading edges of blades by wind carrying sand, which increases roughness and decreases aerodynamic performance. The particle erosion resistance of fiber-reinforced polymers is poor when compared to metallic materials and elastomers, and needs to be improved. It has been shown that the replacement of glass fiber with CNF on the composite surface greatly improves erosion resistance. CNF's have also been shown to provide good electrical conductivity (important for lightning strikes), high damping ratio, and good impact-friction resistance. These properties make CNF-based nanopaper a prospective coating for wind turbine blades.

Tower

Tower Height

Wind velocities increase at higher altitudes due to surface aerodynamic drag (by land or water surfaces) and the viscosity of the air. The variation in velocity with altitude, called wind shear, is most dramatic near the surface. Typically, the variation follows the wind profile power law, which predicts that wind speed rises proportionally to the seventh root of altitude. Doubling the altitude of a turbine, then, increases the expected wind speeds by 10% and the expected power by 34%. To avoid buckling, doubling the tower height generally requires doubling the diameter of the tower as well, increasing the amount of material by a factor of at least four.

At night time, or when the atmosphere becomes stable, wind speed close to the ground usually subsides whereas at turbine hub altitude it does not decrease that much or may even increase. As a result, the wind speed is higher and a turbine will produce more power than expected from the 1/7 power law: doubling the altitude may increase wind speed by 20% to 60%. A stable atmosphere is caused by radiative cooling of the surface and is common in a temperate climate: it usually occurs when there is a (partly) clear sky at night. When the (high altitude) wind is strong (a 10-meter wind speed higher than approximately 6 to 7 m/s) the stable atmosphere is disrupted because of friction turbulence and the atmosphere will turn neutral. A daytime atmosphere is either neutral (no net radiation; usually with strong winds and heavy clouding) or unstable (rising air because of ground heating—by the sun). Here again the 1/7 power law applies or is at least a good approximation of the wind profile. Indiana had been rated as having a wind capacity of 30,000 MW, but by raising the expected turbine height from 50 m to 70 m, the wind capacity estimate was raised to 40,000 MW, and could be double that at 100 m.

For HAWTs, tower heights approximately two to three times the blade length have been found to balance material costs of the tower against better utilisation of the more expensive active components.

Sections of a wind turbine tower, transported in a bulk carrier ship

Road size restrictions makes transportation of towers with a diameter of more than 4.3 m difficult. Swedish analyses show that it is important to have the bottom wing tip at least 30 m above the tree tops, but a taller tower requires a larger tower diameter. A 3 MW turbine may increase output from 5,000 MWh to 7,700 MWh per year by going from 80 to 125 meter tower height. A tower profile made of connected shells rather than cylinders can have a larger diameter and still be transportable. A 100 m prototype tower with TC bolted 18 mm 'plank' shells has been erected at the wind turbine test center Høvsøre in Denmark and certified by Det Norske Veritas, with a Siemens nacelle. Shell elements can be shipped in standard 12 m shipping containers, and 2½ towers per week are produced this way.

As of 2003, typical modern wind turbine installations use towers about 210 ft (65 m) high. Height is typically limited by the availability of cranes. This has led to a variety of proposals for "partially self-erecting wind turbines" that, for a given available crane, allow taller towers that put a turbine in stronger and steadier winds, and "self-erecting wind turbines" that can be installed without cranes.

Tower Materials

Currently, the majority of wind turbines are supported by conical tubular steel towers. These towers represent 30% – 65% of the turbine weight and therefore account for a large percentage of the turbine transportation costs. The use of lighter materials in the tower could greatly reduce the overall transport and construction cost of wind turbines, however the stability must be maintained. Higher grade S500 steel costs 20%-25% more than S335 steel (standard structural steel), but it requires 30% less material because of its improved strength. Therefore, replacing wind turbine towers with S500 steel would result in a net savings in both weight and cost.

Another disadvantage of conical steel towers is that constructing towers that meet the requirements of wind turbines taller than 90 meters proves challenging. High performance concrete shows potential to increase tower height and increase the lifetime of the towers. A hybrid of prestressed concrete and steel has shown improved performance over standard tubular steel at tower heights of 120 meters. Concrete also gives the benefit of allowing for small precast sections to be assembled on site, avoiding the challenges steel faces during transportation. One downside of concrete towers is the higher CO_2 emissions during concrete production as compared to steel. However, the overall environmental benefit should be higher if concrete towers can double the wind turbine lifetime.

Wood is being investigated as a material for wind turbine towers, and a 100 metre tall tower supporting a 1.5 MW turbine has been erected in Germany. The wood tower shares the same transportation benefits of the segmented steel shell tower, but without the steel resource consumption.

Connection to the Electric Grid

All grid-connected wind turbines, from the first one in 1939 until the development of variable-speed grid-connected wind turbines in the 1970s, were fixed-speed wind turbines. As recently as 2003, nearly all grid-connected wind turbines operated at exactly constant speed (synchronous generators) or within a few percent of constant speed (induction generators). As of 2011, many operational wind turbines used fixed speed induction generators (FSIG). As of 2011, most new grid-connected wind turbines are variable speed wind turbines—they are in some variable speed configuration.

Early wind turbine control systems were designed for peak power extraction, also called maximum power point tracking—they attempt to pull the maximum possible electrical power from a given wind turbine under the current wind conditions. More recent wind turbine control systems deliberately pull less electrical power than they possibly could in most circumstances, in order to provide other benefits, which include:

- spinning reserves to quickly produce more power when needed—such as when some other generator suddenly drops from the grid—up to the max power supported by the current wind conditions.

- Variable-speed wind turbines can (very briefly) produce more power than the current wind conditions can support, by storing some wind energy as kinetic energy (accelerating during brief gusts of faster wind) and later converting that kinetic energy to electric energy (decelerating, either when more power is needed elsewhere, or during short lulls in the wind, or both).

- damping (electrical) subsynchronous resonances in the grid

- damping (mechanical) resonances in the tower

The generator in a wind turbine produces alternating current (AC) electricity. Some turbines drive an AC/AC converter—which converts the AC to direct current (DC) with a rectifier and then back to AC with an inverter—in order to match the frequency and phase of the grid. However, the most common method in large modern turbines is to instead use a doubly fed induction generator directly connected to the electricity grid.

A useful technique to connect a permanent magnet synchronous generator to the grid is by using a back-to-back converter. Also, we can have control schemes so as to achieve unity power factor in the connection to the grid. In that way the wind turbine will not consume reactive power, which is the most common problem with wind turbines that use induction machines. This leads to a more stable power system. Moreover, with different control schemes a wind turbine with a permanent magnet synchronous generator can provide or consume reactive power. So, it can work as a dynamic capacitor/inductor bank so as to help with the power systems' stability.

Below we show the control scheme so as to achieve unity power factor :

Reactive power regulation consists of one PI controller in order to achieve operation with unity power factor (i.e. $Q_{grid} = 0$). It is obvious that I_{dN} has to be regulated to reach zero at steady-state ($I_{dNref} = 0$).

Foundations

Wind turbine foundations

Wind turbines, by their nature, are very tall slender structures, this can cause a number of issues when the structural design of the foundations are considered.

The foundations for a conventional engineering structure are designed mainly to transfer the vertical load (dead weight) to the ground, this generally allows for a comparatively unsophisticated arrangement to be used. However, in the case of wind turbines, due to the high wind and environmental loads experienced there is a significant horizontal dynamic load that needs to be appropriately restrained.

This loading regime causes large moment loads to be applied to the foundations of a wind turbine. As a result, considerable attention needs to be given when designing the footings to ensure that the turbines are sufficiently restrained to operate efficiently. In the current Det Norske Veritas (DNV) guidelines for the design of wind turbines the angular deflection of the foundations are limited to 0.5°. DNV guidelines regarding earthquakes suggest that horizontal loads are larger than vertical loads for offshore wind turbines, while guidelines for tsunamis only suggest designing for maximum sea waves. In contrast, IEC suggests considering tsunami loads.

Scale model tests using a 50-g centrifuge are being performed at the Technical University of Denmark to test monopile foundations for offshore wind turbines at 30 to 50-m water depth.

Costs

Liftra Blade Dragon installing a single blade on wind turbine hub.

The modern wind turbine is a complex and integrated system. Structural elements comprise the majority of the weight and cost. All parts of the structure must be inexpensive, lightweight, durable, and manufacturable, under variable loading and environmental conditions. Turbine systems that have fewer failures, require less maintenance, are lighter and last longer will lead to reducing the cost of wind energy.

One way to achieve this is to implement well-documented, validated analysis codes, according to a 2011 report from a coalition of researchers from universities, industry, and government, supported by the Atkinson Center for a Sustainable Future.

The major parts of a modern turbine may cost (percentage of total): tower 22%, blades 18%, gearbox 14%, generator 8%.

Efficiency and Wind Speed

The efficiency of a wind turbine is a maximum at its design wind velocity, and efficiency decreases with the fluctuations in wind. The lowest velocity at which the turbine develops its full power is known as rated wind velocity. Below some minimum wind velocity, no useful power output can be produced from wind turbine. There are limits on both the minimum and maximum wind velocity for the efficient operation of wind turbines.

Design Specification

The design specification for a wind-turbine will contain a power curve and guaranteed availability. With the data from the wind resource assessment it is possible to calculate commercial viability. The typical operating temperature range is −20 to 40 °C (−4 to 104 °F). In areas with extreme climate (like Inner Mongolia or Rajasthan) specific cold and hot weather versions are required.

Wind turbines can be designed and validated according to IEC 61400 standards.

Low Temperature

Utility-scale wind turbine generators have minimum temperature operating limits which apply in areas that experience temperatures below −20 °C. Wind turbines must be protected from ice accumulation. It can make anemometer readings inaccurate and which, in certain turbine control designs, can cause high structure loads and damage. Some turbine manufacturers offer low-temperature packages at a few percent extra cost, which include internal heaters, different lubricants, and different alloys for structural elements. If the low-temperature interval is combined with a low-wind condition, the wind turbine will require an external supply of power, equivalent to a few percent of its rated power, for internal heating. For example, the St. Leon, Manitoba project has a total rating of 99 MW and is estimated to need up to 3 MW (around 3% of capacity) of station service power a few days a year for temperatures down to −30 °C. This factor affects the economics of wind turbine operation in cold climates.

Windmill

The smock mill Goliath in front of the wind farm Growind in Eemshaven in the Netherlands

A windmill is a mill that converts the energy of wind into rotational energy by means of vanes called sails or blades. Centuries ago, windmills usually were used to mill grain, pump water, or both. Thus they often were gristmills, windpumps, or both. The majority of modern windmills take the form of wind turbines used to generate electricity, or windpumps used to pump water, either for land drainage or to extract groundwater.

Windmills in Antiquity

The windwheel of the Greek engineer Heron of Alexandria in the first century is the earliest known instance of using a wind-driven wheel to power a machine. Another early example of a wind-driven wheel was the prayer wheel, which has been used in Tibet and China since the fourth century. It has been claimed that the Babylonian emperor Hammurabi planned to use wind power for his ambitious irrigation project in the seventeenth century BCE.

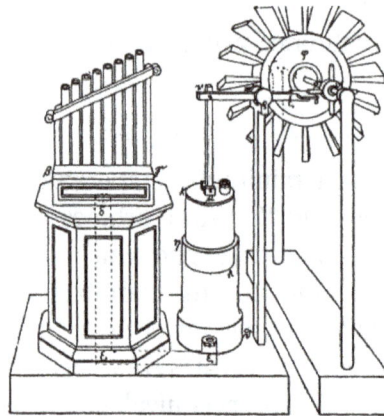

Heron's wind-powered organ

Horizontal Windmills

The first practical windmills had sails that rotated in a horizontal plane, around a vertical axis. According to Ahmad Y. al-Hassan, these panemone windmills were invented in eastern Persia as recorded by the Persian geographer Estakhri in the ninth century. The authenticity of an earlier anecdote of a windmill involving the second caliph Umar (AD 634–644) is questioned on the grounds that it appears in a tenth-century document. Made of six to 12 sails covered in reed matting or cloth material, these windmills were used to grind grain or draw up water, and were quite different from the later European vertical windmills. Windmills were in widespread use across the Middle East and Central Asia, and later spread to China and India from there.

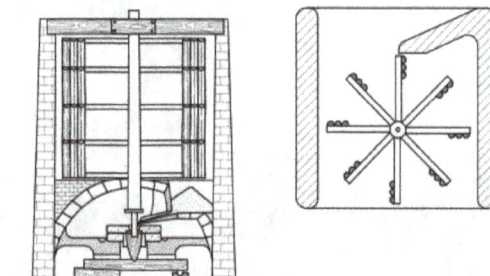

The Persian horizontal windmill

A similar type of horizontal windmill with rectangular blades, used for irrigation, can also be found in thirteenth-century China (during the Jurchen Jin Dynasty in the north), introduced by the travels of Yelü Chucai to Turkestan in 1219.

Hooper's Mill, Margate, Kent, an eighteenth-century European horizontal windmill

Horizontal windmills were built, in small numbers, in Europe during the 18th and nineteenth centuries, for example Fowler's Mill at Battersea in London, and Hooper's Mill at Margate in Kent. These early modern examples seem not to have been directly influenced by the horizontal windmills of the Middle and Far East, but to have been independent inventions by engineers influenced by the Industrial Revolution.

Vertical Windmills

Due to a lack of evidence, debate occurs among historians as to whether or not Middle Eastern horizontal windmills triggered the original development of European windmills. In northwestern Europe, the horizontal-axis or vertical windmill (so called due to the plane of the movement of its sails) is believed to date from the last quarter of the twelfth century in the triangle of northern France, eastern England and Flanders.

The earliest certain reference to a windmill in Europe (assumed to have been of the vertical type) dates from 1185, in the former village of Weedley in Yorkshire which was located at the southern tip of the Wold overlooking the Humber estuary. A number of earlier, but less certainly dated, twelfth-century European sources referring to windmills have also been found. These earliest mills were used to grind cereals.

Post Mill

The evidence at present is that the earliest type of European windmill was the post mill, so named because of the large upright post on which the mill's main structure (the "body" or "buck") is balanced. By mounting the body this way, the mill is able to rotate to face the wind direction; an essential requirement for windmills to operate economically in north-western Europe, where wind directions are variable. The body contains all the milling machinery. The first post mills were of the sunken type, where the post was buried in an earth mound to support it. Later, a wooden support was developed called the trestle. This was often covered over or surrounded by a roundhouse to protect the trestle from the weather and to provide storage space. This type of windmill was the most common in Europe until the nineteenth century, when more powerful tower and smock mills replaced them.

Hollow-Post Mill

In a hollow-post mill, the post on which the body is mounted is hollowed out, to accommodate the drive shaft. This makes it possible to drive machinery below or outside the body while still being able to rotate the body into the wind. Hollow-post mills driving scoop wheels were used in the Netherlands to drain wetlands from the fourteenth century onwards.

Tower Mill

By the end of the thirteenth century, the masonry tower mill, on which only the cap is rotated rather than the whole body of the mill, had been introduced. The spread of tower mills came with a growing economy that called for larger and more stable sources of power, though they were more expensive to build. In contrast to the post mill, only the cap of the tower mill needs to be turned into the wind, so the main structure can be made much taller, allowing the sails to be made longer, which enables them to provide useful work even in low winds. The cap can be turned into the wind either by winches or gearing inside the cap or from a winch on the tail pole outside the mill. A method of keeping the cap and sails into the wind automatically is by using a fantail, a small windmill mounted at right angles to the sails, at the rear of the windmill. These are also fitted to tail poles of post mills and are common in Great Britain and English-speaking countries of the former British Empire, Denmark, and Germany but rare in other places. Around some parts of the Mediterranean Sea, tower mills with fixed caps were built because the wind's direction varied little most of the time.

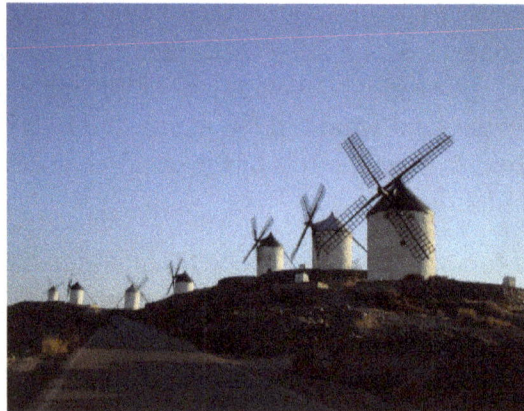

Tower mills in Spain

Smock Mill

The smock mill is a later development of the tower mill, where the tower is replaced by a wooden framework, called the "smock." The smock is commonly of octagonal plan, though examples with more, or fewer, sides exist. The smock is thatched, boarded or covered by other materials, such as slate, sheet metal, or tar paper. The lighter construction in comparison to tower mills make smock mills practical as drainage mills as these often had to be built in areas with unstable subsoil. Having originated as a drainage mill, smock mills are also used for a variety of purposes. When used in a built-up area it is often placed on a masonry base to raise it above the surrounding buildings.

Two smock mills with a stage in Greetsiel, Germany

Mechanics

Sails

Common sails consist of a lattice framework on which a sailcloth is spread. The miller can adjust the amount of cloth spread according to the amount of wind available and power needed. In medieval mills, the sailcloth was wound in and out of a ladder type arrangement of sails. Postmedieval mill sails had a lattice framework over which the sailcloth was spread, while in colder climates, the cloth was replaced by wooden slats, which were easier to handle in freezing conditions. The jib sail is commonly found in Mediterranean countries, and consists of a simple triangle of cloth wound round a spar.

Windmill in Kuremaa, Estonia

In all cases, the mill needs to be stopped to adjust the sails. Inventions in Great Britain in the late eighteenth and nineteenth centuries led to sails that automatically adjust to the wind speed without the need for the miller to intervene, culminating in patent sails invented by William Cubitt in 1807. In these sails, the cloth is replaced by a mechanism of connected shutters.

In France, Pierre-Théophile Berton invented a system consisting of longitudinal wooden slats connected by a mechanism that lets the miller open them while the mill is turning. In the twentieth century, increased knowledge of aerodynamics from the development of the airplane led to further improvements in efficiency by German engineer Bilau and several Dutch millwrights.

The majority of windmills have four sails. Multiple-sailed mills, with five, six or eight sails, were built in Great Britain (especially in and around the counties of Lincolnshire and Yorkshire), Germany, and less commonly elsewhere. Earlier multiple-sailed mills are found in Spain, Portugal, Greece, parts of Romania, Bulgaria, and Russia. A mill with an even number of sails has the advantage of being able to run with a damaged sail and the one opposite removed without resulting in an unbalanced mill.

De Valk windmill in mourning position following the death of Queen Wilhelmina of the Netherlands in 1962

In the Netherlands the stationary position of the sails, i.e. when the mill is not working, has long been used to give signals. A slight tilt of the sails before the main building signals joy, while a tilt after the building signals mourning. Across the Netherlands, windmills were placed in mourning position in honor of the Dutch victims of the 2014 Malaysian Airlines Flight 17 shootdown.

Machinery

Littlefield, Texas, claims the world's tallest windmill.

Gears inside a windmill convey power from the rotary motion of the sails to a mechanical device. The sails are carried on the horizontal windshaft. Windshafts can be wholly made of wood, or wood with a cast iron poll end (where the sails are mounted) or entirely of cast iron. The brake wheel is fitted onto the windshaft between the front and rear bearing. It has the brake around the outside of the rim and teeth in the side of the rim which drive the horizontal gearwheel called wallower on the top end of the vertical upright shaft. In grist mills, the great spur wheel, lower down the upright shaft, drives one or more stone nuts on the shafts driving each millstone. Post mills sometimes have a head and/or tail wheel driving the stone nuts directly, instead of the spur gear arrangement. Additional gear wheels drive a sack hoist or other machinery. The machinery differs if the windmill is used for other applications than milling grain. A drainage mill uses another set of gear wheels on the bottom end of the upright shaft to drive a scoop wheel or Archimedes' screw. Sawmills use a crankshaft to provide a reciprocating motion to the saws. Windmills have been used to power many other industrial processes, including papermills, threshing mills, and to process oil seeds, wool, paints and stone products.

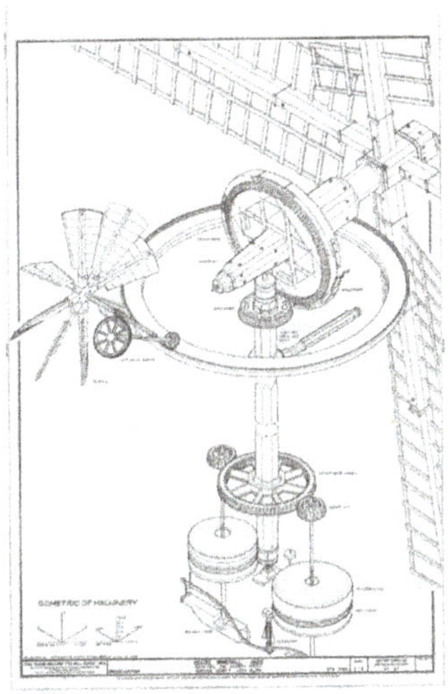

An isometric drawing of the machinery of the Beebe Windmill

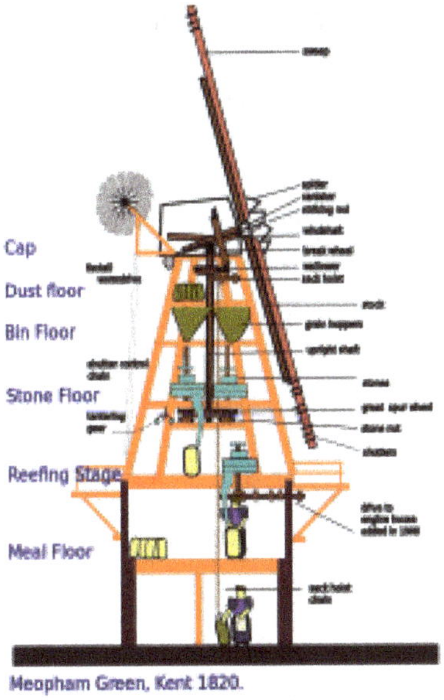

Cap
Dust floor
Bin Floor
Stone Floor
Reefing Stage
Meal Floor

Meopham Green, Kent 1820.

Diagram of the smock mill at Meopham,

Spread and Decline

The total number of wind-powered mills in Europe is estimated to have been around 200,000 at its peak, which is modest compared to some 500,000 waterwheels. Windmills were applied in regions where there was too little water, where rivers freeze in winter and in flat lands where the flow of the river was too slow to provide the required power. With the coming of the industrial revolution, the importance of wind and water as primary industrial energy sources declined and were eventually replaced by steam (in steam mills) and internal combustion engines, although windmills continued to be built in large numbers until late in the nineteenth century. More recently, windmills have been preserved for their historic value, in some cases as static exhibits when the antique machinery is too fragile to put in motion, and in other cases as fully working mills.

A windmill in Wales, United Kingdom. 1815.

Of the 10,000 windmills in use in the Netherlands around 1850, about 1000 are still standing. Most of these are being run by volunteers, though some grist mills are still operating commercially. Many of the drainage mills have been appointed as backup to the modern pumping stations. The Zaan district has been said to have been the first industrialized region of the world with around 600 operating wind-powered industries by the end of the eighteenth century. Economic fluctuations and the industrial revolution had a much greater impact on these industries than on grain and drainage mills, so only very few are left.

Oilmill De Zoeker, paintmill De Kat and paltrok sawmill De Gekroonde Poelenburg at the Zaanse Schans

Construction of mills spread to the Cape Colony in the seventeenth century. The early tower mills did not survive the gales of the Cape Peninsula, so in 1717, the Heeren XVII sent carpenters, masons, and materials to construct a durable mill. The mill, completed in 1718, became known as the *Oude Molen* and was located between Pinelands Station and the Black River. Long since demolished, its name lives on as that of a Technical school in Pinelands. By 1863, Cape Town could boast 11 mills stretching from Paarden Eiland to Mowbray.

Don Quijote being struck by a windmill, illustration by Paul Gustave Louis Christophe Doré.

Wind Turbines

A wind turbine is a windmill-like structure specifically developed to generate electricity. They can be seen as the next step in the development of the windmill. The first wind turbines were built by the end of the nineteenth century by Prof James Blyth in Scotland (1887), Charles F. Brush in Cleveland, Ohio (1887–1888) and Poul la Cour in Denmark (1890s). La Cour's mill from 1896 later became the local powerplant of the village Askov. By 1908 there were 72 wind-driven electric generators in Denmark, ranging from 5 to 25 kW. By the 1930s, windmills were widely used to generate electricity on farms in the United States where distribution systems had not yet been installed, built by companies such as Jacobs Wind, Wincharger, Miller Airlite, Universal Aeroelectric, Paris-Dunn, Airline, and Winpower. The Dunlite Corporation produced turbines for similar locations in Australia.

Rønland Windpark in Denmark

Forerunners of modern horizontal-axis utility-scale wind generators were the WIME-3D in service in Balaklava USSR from 1931 until 1942, a 100-kW generator on a 30-m (100-ft) tower, the Smith-Putnam wind turbine built in 1941 on the mountain known as Grandpa's Knob in Castleton, Vermont, United States of 1.25 MW and the NASA wind turbines developed from

1974 through the mid-1980s. The development of these 13 experimental wind turbines pioneered many of the wind turbine design technologies in use today, including: steel tube towers, variable-speed generators, composite blade materials, and partial-span pitch control, as well as aerodynamic, structural, and acoustic engineering design capabilities. The modern wind power industry began in 1979 with the serial production of wind turbines by Danish manufacturers Kuriant, Vestas, Nordtank, and Bonus. These early turbines were small by today's standards, with capacities of 20–30 kW each. Since then, commercial turbines have increased greatly in size, with the Enercon E-126 capable of delivering up to 7 MW, while wind turbine production has expanded to many countries.

A group of wind turbines in Zhangjiakou, China

As the 21st century began, rising concerns over energy security, global warming, and eventual fossil fuel depletion led to an expansion of interest in all available forms of renewable energy. Worldwide, many thousands of wind turbines are now operating, with a total nameplate capacity of 194,400 MW. Europe accounted for 48% of the total in 2009.

Windpumps

Windpumps were used to pump water since at least the 9th century in what is now Afghanistan, Iran and Pakistan. The use of wind pumps became widespread across the Muslim world and later spread to China and India. Windmills were later used extensively in Europe, particularly in the Netherlands and the East Anglia area of Great Britain, from the late Middle Ages onwards, to drain land for agricultural or building purposes.

The American windmill, or wind engine, was invented by Daniel Halladay in 1854 and was used mostly for lifting water from wells. Larger versions were also used for tasks such as sawing wood, chopping hay, and shelling and grinding grain. In early California and some other states, the windmill was part of a self-contained domestic water system which included a hand-dug well and a wooden water tower supporting a redwood tank enclosed by wooden siding known as a tankhouse. During the late 19th century steel blades and steel towers replaced wooden construction. At their peak in 1930, an estimated 600,000 units were in use. Firms such as U.S. Wind Engine and Pump Company, Challenge Wind Mill and Feed Mill Company, Appleton Manufacturing Company, Star, Eclipse, Fairbanks-Morse, and Aermotor became the main suppliers in North and South America. These windpumps are used extensively on farms and ranches in the United States, Canada, Southern Africa, and Australia. They feature a large number of blades, so they turn slowly with considerable torque in low winds and are self-regulating in high winds. A tower-top gearbox and crankshaft convert the rotary motion into reciprocating strokes carried downward through a rod to the pump cylinder below. Such mills pumped water and powered feed mills, saw mills, and agricultural machinery.

Aermotor-style windpump in South Dakota, USA

In Australia, the Griffiths Brothers at Toowoomba manufactured windmills from 1876, with the trade name Southern Cross Windmills in use from 1903. These became an icon of the Australian rural sector by utilizing the water of the Great Artesian Basin.

Sail

A sail is means for redirecting the power of the wind to propel a craft on water, ice or land. In doing so, sails mobilize lifting properties as air passes along the surface and they mobilize drag properties to the degree that air is directed at the surface. When both lift and drag are present, they function similarly to a wing in a vertical orientation. In most cases sails are supported by a mast rigidly attached to the sailing craft, however some craft employ a flexible mount for a mast. Sails also employ spars and battens to determine shape in the axis perpendicular to the mast. As a result, sails come in a variety of shapes that include both triangular and quadrilateral configurations, usually with curved edges that promote curvature of the sail.

Kites that are used to propel certain sailing craft are differentiated from sails in that they are supported and controlled by lines that lead from the kite to the craft.

A lateen sail

History of Sails

Archaeological studies of the Cucuteni-Trypillian culture ceramics show use of sailing boats from the sixth millennium onwards. Excavations of the Ubaid period (c. 6000 -4300 BC) in Mesopotamia provides direct evidence of sailing boats. Sails from ancient Egypt are depicted around 3200 BCE, where reed boats sailed upstream against the River Nile's current. Ancient Sumerians used square rigged sailing boats at about the same time, and it is believed they established sea trading routes as far away as the Indus valley. The proto-Austronesian words for sail, *lay(r)*, and other rigging parts date to about 3000 BCE when this group began their Pacific expansion. Greeks and Phoenicians began trading by ship by around 1,200 BCE.

Yacht sails seen from the deck

Square sails mounted on yardarms perpendicular to the boat's hull are very good for downwind sailing; they dominated in the ancient Mediterranean and spread to Northern Europe, and were independently invented in China and Ecuador. Although fore-and-aft rigs have become more popular on modern yachts, square sails continue to power full-rigged ships through the Age of Sail and to the present day. Triangular fore-and-aft rigs were invented in the Mediterranean as single yarded lateen sails and independently in the Pacific as the more efficient bi sparred crab claw sail, and continue to be used throughout the world. During the 16th-19th centuries other fore-and-aft sails were developed in Europe, such as the spritsail, gaff rig, jib/genoa/staysail, and Bermuda rig, improving European upwind sailing ability.

In an interesting recent development, an elderly trawler, TS *Pelican*, was fitted with what are thought to have been the unorthodox riggings used by the Barbary pirates in the 16th century. The resultant performance has been remarkable, with the Pelican sailing, at speed, over 20 degrees nearer the wind than any square rigger.

Space satellites have successfully deployed solar sails which use radiation pressure or solar wind to propel them.

Use of Sails

Sails are primarily used on the water by sailing ships and sail boats as a propulsion system. For purposes of commerce, sails have been greatly superseded by other forms of propulsion, such as the internal combustion engine. For recreation, however, sailing vessels remain popular. The most familiar type of sailboat, a small pleasure yacht, usually has a sail-plan called a sloop. This has two sails in a fore-and-aft arrangement: the mainsail and the jib.

The mainsail extends aftward (backwards) and is secured the whole length of its front edge to the mast and its aft corner (clew) to a boom, generally, also hung from the mast. The sails of tall ships are attached to wooden timbers or "spars".

The jib is secured along its leading edge to a forestay (strong wire) strung from the top of the mast to the bowsprit on the bow (nose) of the boat. A genoa is also used on some boats. It is a type of jib that is large enough to overlap the mainsail, and cut so that it is fuller than an ordinary jib producing a greater driving force in lighter winds.

A spinnaker is also used on some boats to help move the sailboat faster downwind. The spinnaker is often a colourful sail and can be either symmetrical or asymmetrical.

Fore-and-aft sails can be switched from one side of the boat to the other in order to provide propulsion as the sailboat changes direction relative to the wind. When the boat's stern crosses the wind, this is called gybing; when the bow crosses the wind, it is called tacking. Tacking repeatedly from port to starboard and/or vice versa, called "beating", is done in order to allow the boat to follow a course into the wind. Modern boats can sail as close as 30 degrees to the wind.

A primary feature of a properly designed sail is an amount of "draft", caused by curvature of the surface of the sail. When the leading edge of a sail is oriented into the wind, the correct curvature helps maximise lift while minimising turbulence and drag, much like the carefully designed curves of aircraft wings. Modern sails are manufactured with a combination of broadseaming and non-stretch fabric. The former adds draft, while the latter allows the sail to keep a constant shape as the wind pressure increases. The draft of the sail can be reduced in stronger winds by use of a cunningham and outhaul, and also by bending the mast and increasing the downward pressure of the boom by use of a boom vang.

Other sail powered machines include ice yachts, windmills, kites, signs, hang gliders, electric generators, windsurfers, and land sailing vehicles.

Sail construction is governed by the science of aerodynamics.

Types of Rig

Generally speaking, sailing vessels employ two main types of rig: the square rig and the fore-and-aft rig.

The square rig, which reached its maximum development in the clipper ships and trading barques of the late 19th and early 20th century, relies on rectangular sails hung beneath yards, themselves suspended from the masts and set "square" (i.e., at a right angle to) the keel of the ship. This kind of rig requires an enormous amount of rigging (at least nine ropes per sail) and cannot sail closer than about 60° to the wind. Few vessels of this type are seen today, other than the spectacular ones used for sail training. Most square rigged vessels also carry at least some fore-and-aft sails.

Ship Garthsnaid at sea c. 1920. Men can be seen in the rigging.

The fore-and-aft rig is far more common: nearly every dinghy and yacht uses this type of rig, in which the sails are mounted parallel to the keel and are secured to the fore of the ship and to the aft rather than side to side. A large mainsail is often rigged abaft the mast(s) and usually a jib in front of it. The foot of the mainsail is usually extended by a boom. Each sail needs only two or three ropes for its basic control.

Sail Aerodynamics

Sails propel the boat in one of two ways. When the boat is going in the direction of the wind, the sails may be set merely to trap the air as it flows by. Sails acting in this way are aerodynamically stalled. Drag, always parallel to the wind, contributes the predominant driving force.

The other way sails propel the boat occurs when the boat is traveling across or into the wind. The sails acting as airfoils propel the boat by redirecting the wind coming in from the side towards the rear. By the law of conservation of momentum, the wind moves the sail as the sail redirects downwash air backwards. Air pressure differences across the sail area result in forces on sails including drag and lift. A component of the lift is the main driving force.

The sails can also act as airfoils in some downwind situations, e.g. spinnakers and square-rigged sails can be trimmed so that their *upper* edges become leading edges and they operate as airfoils again, but with airflow directed more or less vertically downwards. This mode of trim also provides the boat with some actual lift and may reduce both wetted area and the risk of 'digging into' waves. In stronger winds, turbulence created behind stalled sails can lead to aerodynamic instability, which in turn can manifest as increased downwind rolling of the boat.

Sails are often equipped with lightweight tapes or strands (tell-tales) to indicate the airflow in their area. They may be on both sides near the leading edges of the sail, or at the trailing edge of the sail. Horizontal strips sewn into fore-and-aft sails and V-shaped markings on spinnakers assist with judging their shape from on deck. These may even glow in the dark, using luminous tapes.

On a sailing boat, a keel or centreboard helps to prevent the boat from moving sideways. The shape of the keel has a much smaller cross section in the fore and aft axis and a much larger cross section on the athwart axis (across the beam of the boat). The resistance to motion along the smallest cross section is low while resistance to motion across the large cross section is high, so the boat moves forward rather than sideways. In other words, it is easier for the sail to push the boat forward rather than sideways. However, there is always a small amount of sideways motion, or leeway. The keel or centreboard acts as a secondary foil, symmetrically aligned under the vessel front to back, the sidewise forces induced by the sail create an asymmetrical water flow across the foil resulting in opposing lift.

Forces across the boat are resolved by balancing the sideways force from the sail with the sideways resistance of the keel or centreboard. Also, if the boat heels, there are restoring forces due to the shape of the hull and the mass of the ballast in the keel being raised against gravity. Forward forces are balanced by velocity through the water and friction between the hull, keel and the water.

Parts of the Sail

The lower edge of a triangular sail is called the "foot" of the sail, while the upper point is known as the "head". The lower two points of the sail, on either end of the foot, are called the "tack" (forward) and "clew" (aft). The forward edge of the sail is called the "luff" (from which derives the term "luffing", a rippling of the sail when the angle of the wind fails to maintain a good aerodynamic shape near the luff). The aft edge of a sail is called the "leech". The curved sail area beyond a straight line from the head to the clew is known as the "roach". Typically this is greater in a racing sail and may be absent in a cruising sail. The roach is held in shape by sail battens which maybe full length or short.

Modern sails are designed so that either the warp or weft of the cloth is perpendicular to the leech. This places the most elastic axis of the cloth on the luff and foot, where bias stretch can be controlled with folded cloth tabling or rope. Varying tension on the luff and foot with winches, downhauls, or outhauls allows the sailor to adjust the draft to suit wind conditions.

Often tell-tales, small pieces of yarn, are attached to the sail. They are used as a guide when trimming the sail as they indicate the wind flow across the sail.

An alternative approach to sail design is that used in junks, originally an oriental design. It uses horizontal sail curving to produce an efficient and easily controlled sail-plan.

Sail Types

Modern sails can be classified into three main categories:

- Mainsail,

- Headsail also known as the jib sail,

- and Spinnaker or downwind sail (also termed Kite). Special-purpose sails are often a variation of the three main categories.

High-performance yachts, in particular some catamarans such as the International C-Class Catamaran, have used or use rigid wing sails, which are said to provide better performance than traditional soft sails. In particular, a rigid wing sail was used by Stars and Stripes, the defender which won the 1988 America's Cup, and by USA-17, the challenger which won the 2010 America's Cup.

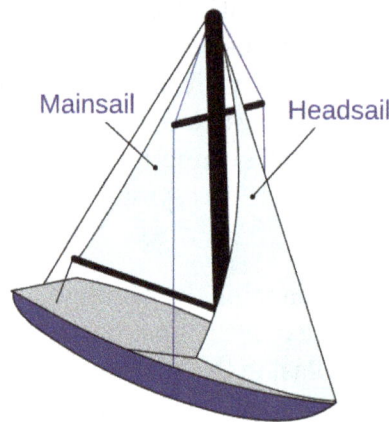

Diagram of Sailboat, in this case a typical monohull sloop with a Bermuda or marconi rig.

Most modern yachts, including bermuda rig, ketch and yawl boats, have a sail "inventory" which usually includes more than one of these types of sails. Although the mainsail is "permanently" hoisted while sailing, headsails and spinnakers can be changed depending on the particular weather conditions to allow better handling and speed.

Mainsails, as the name implies, are the main element of the sailplan. A "motor" as well as a rudder for the boat, mainsails can be as simple as a traditional triangle-shaped, cross-cut sail (see Sail Construction). In most cases, the mainsail isn't changed while sailing, although there are mechanisms to reduce its surface if the wind is very strong (a technique called reefing). In extreme weather, a mainsail can be folded and a trysail hoisted to allow steerage without endangering the boat.

Headsails are the main driving sails when going upwind (sailing towards the wind). There are many types of headsails with Genoa and Jib being the most commonly used. Both these types have different subtypes depending on their intended use. Headsails are usually classified according to their weight (that is, the relative weight of the sailcloth used) and size or total area of the sail. A common classification is numbering from 1 to 3 (larger to smaller) with a description of the use for example: #1 Heavy or #1 Medium/Light. Special types of headsails include the Gennaker (also named Code 0 by some sailmakers), the drifter (a type of Genoa that is used like an asymmetrical spinnaker), the screecher (essentially a large Genoa), the windseeker and storm jib. Certain Genoas and Jibs also have battens which assist in maintaining an optimal shape for the sail.

Spinnakers are used for reaching and running (downwind sailing). They are very light and have a balloon-like shape. As with headsails, there are many types of spinnakers depending on the shape, area and cloth weight. Symmetrical spinnakers are most efficient on runs and dead runs (sailing with wind coming directly from behind) while asymmetric spinnakers are very efficient in reaching (the wind coming from the rear but at an angle to the boat or from the side).

Sail Construction

A sail might look flat when lying on the floor but once it is hoisted, it becomes a three-dimensional curved surface, in essence an airfoil. In order for a sail to be "built", it has to be designed in a number of elements (or panels) which are cut and sewn together to form the foil. In older days, this was rightfully considered an art which was later complemented (and arguably overshadowed) by technology. With the advent of computers, sail manufacturers were able to model their sails using special computer-aided design (CAD) programs and directly feed the data to very accurate laser plotters/cutters which cut the panels from rolls of sail cloth, replacing the traditional manual process (scissors).

The key features that distinguish a "fast" from a "slow" sail are its shape related to the particular boat and rig and its ability to consistently maintain that shape. These two features rely mostly on the design of the sail (the way that the panels are placed with one another) and the sail cloth used.

The traditional parallel-panel (cross-cut) gave way to more complex (radial) designs where the panels have different shapes for the top, mid, and lower sections of the sail depending on pressure of the air caused by its flow over the sail surface. Again aided by CAD and special modelling software the sailmakers use cloths of different weight, placing heavier cloth panels where there is more stress and lighter cloth where there is less to make savings in weight.

Older fabrics (especially cotton and low budget synthetic), have the tendency to stretch with wind pressure which results in distorted and consequently inefficient sail shapes. Moreover, the cloth itself is heavy which adds to the inefficiency. Synthetic materials such as Nylon and Dacron were followed by advanced sail cloths made from exotic material yarns such as Aramid (e.g. Twaron, Technora or kevlar), carbon fibre, HMPE (e.g. Spectra/Dyneema), Zylon (PBO) and Vectran. These materials were a breakthrough in sail technology as they provided the raw material in the manufacture of low-stretch, low-weight and long-life sail cloths. Manufacturers were able to use different weights of yarn to weave cloths with exceptional properties.

Once the panels are sewn together (often by triple-stitch method), the sailmakers complete the sail by placing the finishing elements such as the leech and foot lines, protective patches in the areas where the sail will scrape against hardware (stanchions, spreaders), steel rings and straps at the tack and clew, cleats, batten pockets (if required) and sail numbers.

Lamination

Woven cloth or ribbons of high tensile fabric inserts can be "sandwiched" between two layers of PET film and placed in special ovens under pressure to bond into a single body, a process called lamination. The inserts provide the strength and the PET film the continuity and wind resistance. An alternative method is to sandwich a sheet of PET film between two layers of woven cloth. The latter process is popular when using cloth with high strength and UV tolerance, but an open weave. In the latter process the cloth protects the more easily torn PET film. A more complex sail may combine the processes.

A light-weather sail generally weighs around 100 gram/m^2 and a rough-weather sail/try-out weighs around 500 gram/m^2 although modern laminated sails can weigh considerably less than this depending on the fibres specified.

Advances in Sail Materials and Manufacture

In addition to advances in the exotic materials and consequent cloths themselves, manufacturers have also progressed the manufacturing process with the creation of glued, molded and laminated sails.

Glued sails are regular paneled sails but instead of sewing the pieces together, the sail maker uses a special, ultra-strong polymer glue which bonds through the use of ultrasound.

In molding, a curved mold is designed and created in the optimum (three dimensional) shape of the sail that the sail maker wants to produce. A film of PET film is placed on the mold and a special gantry hovers over the film laying the yarns based on instructions of a computer that has the model of the sail. Once this is done, a second sheet of PET film is placed on top and the whole mold (with the sail) is placed in a vacuum oven which causes the materials to bond (curing). The result is a smooth sail which is lighter and has a wider effective wind range (the minimum and maximum wind speed that the sail can withstand and be effective).

Environmental Impact of Wind Power

The environmental impact of wind power when compared to the environmental impacts of fossil fuels, is relatively minor. Compared with other low carbon power sources, wind turbines have some of the lowest global warming potential per unit of electrical energy generated. According to the IPCC, in assessments of the life-cycle global warming potential of energy sources, wind turbines have a median value of between 12 and 11 (gCO_{2eq}/kWh) depending on whether off- or onshore turbines are being assessed.

While wind turbine installations may cover a large area, they are compatible with many land uses such as farming and grazing, as only small areas of turbine foundations and infrastructure are made unavailable for use.

Wind turbines generate some noise. At a residential distance of 300 metres (980 ft) this may be around 45 dB, which is slightly louder than a refrigerator. At 1.5 km (1 mi) distance most wind turbines become inaudible. From a fundamental biological point of view, it is known that loud noise increases stress and stress causes diseases. Peer-reviewed research has generally supported the view that when properly sited wind turbines do not affect human health from noise. However when improperly sited, data from the monitoring of two groups of growing geese revealed substantially lower body weights and higher concentrations of a stress hormone in the blood of the first group of geese who were situated 50 meters away compared to a second group which was at a distance of 500 meters from the turbine.

Aesthetic aspects of wind turbines and resulting changes of the visual landscape can be significant. Conflicts arise especially in scenic and heritage protected landscapes. Siting restrictions (such as setbacks) have often been implemented to limit any intrusive environmental impacts.

There are reports of bird and bat mortality at wind turbines as there are around other artificial structures. The scale of the ecological impact may or may not be significant, depending on specific circumstances. Prevention and mitigation of wildlife fatalities, and protection of peat bogs, affect the siting and operation of wind turbines.

Basic Operational Considerations

Net Energy Gain

The energy return on investment (EROI) for wind energy is equal to the cumulative electricity generated divided by the cumulative primary energy required to build and maintain a turbine. According to a meta study, in which all existing studies from 1977 to 2007 were reviewed, the EROI for wind ranges from 5 to 35, with the most common turbines in the range of 2 MW nameplate capacity-rotor diameters of 66 meters. On average the EROI is 16, but including the necessary pump storage systems reduces the EROI for wind remarkably- to an order of magnitude worse than fossil fuels or nuclear power. EROI is strongly proportional to turbine size, and larger late-generation turbines average at the high end of this range, at or above 35. Wind turbine manufacturer Vestas claims that initial energy "pay back" is within about 7–9 months of operation for a 1.65-2.0MW wind power plant under low wind conditions, whereas Siemens Wind Power calculates 5–10 months depending on circumstances.

Pollution & Effects on The Grid

Pollution Costs

Wind power consumes no water for continuing operation, and has near negligible emissions directly related to its electricity production. Wind turbines when isolated from the electric grid produce negligible amounts of carbon dioxide, carbon monoxide, sulfur dioxide, nitrogen dioxide, mercury and radioactive waste when in operation, unlike fossil fuel sources and nuclear energy station fuel production, respectively.

With the construction phase largely to blame, wind turbines emit slightly more particulate matter(PM), a form of air pollution, at an "exception" rate higher per unit of energy generated(kWh) than a fossil gas electricity station("NGCC"), and also emit more heavy metals and PM than nuclear stations, per unit of energy generated. As far as total pollution costs in economic terms, in a comprehensive 2006 European study, alpine Hydropower was found to exhibit the lowest external pollution, or externality, costs of all electricity generating systems, below 0.05 c€/kWh. Wind power externality costs were found to be 0.09 - 0.12c€/kW, while nuclear energy had a 0.19 c€/kWh value and fossil fuels generated 1.6 - 5.8 c€/kWh of downstream costs. With the exception of the latter fossil fuels, these are negligible costs in comparison to the cost of electricity production, which is approximately 10 c€/kWh in European countries.

Findings When Connected to The Grid

A typical study of a wind farms Life cycle assessment, when not connected to the electric grid, usually results in similar findings as the following 2006 analysis of 3 installations in the US Midwest, where the carbon dioxide(CO_2) emissions of wind power ranged from 14 to 33 tonnes (15 to 36 short tons) per GWh(14 - 33 gCO_2/kWh) of energy produced, with most of the CO_2 emission intensity coming from producing the concrete for wind-turbine foundations. By combining similar data from numerous individual studies in a meta-analysis, the median global warming potential for wind power was found to be 11-12g CO2/kWh and unlikely to change significantly.

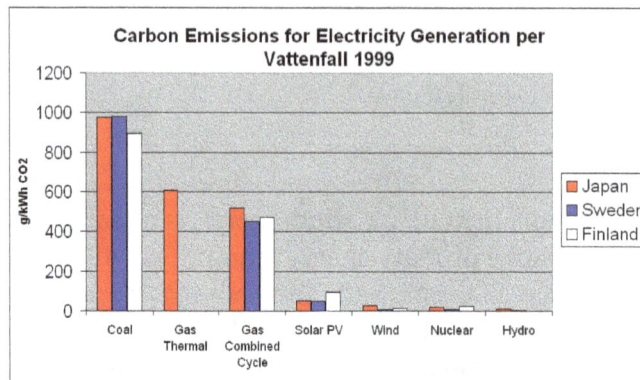

The Vattenfall utility company study found Hydroelectric, nuclear stations and wind turbines to have far less greenhouse emissions than other sources represented.

However these relatively low pollution values begin to increase as greater and greater wind energy is added to the grid, or wind power 'electric grid penetration' levels are reached. Due to the effects of attempting to balance out the energy demands on the grid, from Intermittent power sources e.g. wind power(sources which have low capacity factors due to the weather), this either requires the construction of large energy storage projects, which have their own emission intensity which must be added to wind power's system-wide pollution effects, or it requires more frequent reliance on fossil fuels than the spinning reserve requirements necessary to back up more dependable sources. With the latter combination presently being the more common.

This higher dependence on back-up/Load following power plants to ensure a steady power grid output has the knock-on-effect of more frequent inefficient(in CO_2e g/kWh) throttling up and down of these other power sources in the grid to facilitate the intermittent power source's variable output. When one includes the total effect of intermittent sources on other power sources in the grid system, that is, including these inefficient start up emissions of backup power sources to cater for wind energy, into wind energy's total system wide life cycle, this results in a higher real-world wind energy emission intensity. Higher than the direct g/kWh value that is determined from looking at the power source in isolation and thus ignores all down-stream detrimental/inefficiency effects it has on the grid. This higher dependence on back-up/Load following power plants to ensure a steady power grid output forces fossil power plants to operate in less efficient states. In a 2012 paper that appeared in the *Journal of Industrial Ecology* it states.

"The thermal efficiency of fossil-based power plants is reduced when operated at fluctuating and suboptimal loads to supplement wind power, which may degrade, to a certain extent, the GHG(Greenhouse gas) benefits resulting from the addition of wind to the grid. A study conducted

by Pehnt and colleagues (2008) reports that a moderate level of [grid] wind penetration (12%) would result in efficiency penalties of 3% to 8%, depending on the type of conventional power plant considered. Gross and colleagues (2006) report similar results, with efficiency penalties ranging from nearly 0% to 7% for up to 20% [of grid] wind penetration. Pehnt and colleagues (2008) conclude that the results of adding offshore wind power in Germany on the background power systems maintaining a level supply to the grid and providing enough reserve capacity amount to adding between 20 and 80 g CO2-eq/kWh to the life cycle GHG emissions profile of wind power.'"

In comparison to other low carbon power sources Wind turbines, when assessed in isolation, have a median life cycle emission value of between 11 and 12 (gCO_{2eq}/kWh). The more dependable alpine Hydropower and nuclear stations have median total life cycle emission values of 24 and 12 g CO2-eq/kWh respectively.

While an increase in emissions due to the practical issues of load balancing is an issue, Pehnt et al. still conclude that these 20 and 80 g CO2-eq/kWh added penalties still result in wind being roughly ten times less polluting than fossil gas and coal which emit ~400 and 900 g CO2-eq/kWh respectively.

However, as these losses occur due to cycling of fossil power plants, they become smaller when more renewables are added to the power system and fossil power plants are replaced by renewables. Consequently, also the emissions of the renewables drop with the progressing energy transition.

Rare-Earth Use

The production of permanent magnets used in some wind turbines makes use of neodymium. Primarily exported by China, pollution concerns associated with the extraction of this rare-earth element have prompted government action in recent years, and international research attempts to refine the extraction process. Research is underway on turbine and generator designs which reduce the need for neodymium, or eliminate the use of rare-earth metals altogether. Additionally, the large wind turbine manufacturer Enercon GmbH chose very early not to use permanent magnets for its direct drive turbines, in order to avoid responsibility for the adverse environmental impact of rare earth mining.

Ecology

Land Use

Wind farms are often built on land that has already been impacted by land clearing. The vegetation clearing and ground disturbance required for wind farms is minimal compared with coal mines and coal-fired power stations. If wind farms are decommissioned, the landscape can be returned to its previous condition.

A study by the US National Renewable Energy Laboratory of US wind farms built between 2000 and 2009 found that, on average, only 1.1 percent of the total wind farm area suffered surface disturbance, and only 0.43 percent was permanently disturbed by wind power installations. On average, there were 63 hectares (156 acres) of total wind farm area per MW of capacity, but only 0.27 hectares (0.67 acres) of permanently disturbed area per MW of wind power capacity.

In the UK many prime wind farm sites - locations with the best average wind speeds - are in upland areas which are frequently covered by blanket bog. This type of habitat exists in areas of relatively high rainfall where large areas of land remain permanently sodden. Construction work may create a risk of disruption to peatland hydrology which could cause localised areas of peat within the area of a wind farm to dry out, disintegrate, and so release their stored carbon. At the same time, the warming climate which renewable energy schemes seek to mitigate could itself pose an existential threat to peatlands throughout the UK. A Scottish MEP campaigned for a moratorium on wind developments on peatlands saying that "Damaging the peat causes the release of more carbon dioxide than wind farms save". A 2014 report for the Northern Ireland Environment Agency noted that siting wind turbines on peatland could release considerable carbon dioxide from the peat, and also damage the peatland contributions to flood control and water quality: "The potential knock-on effects of using the peatland resource for wind turbines are considerable and it is arguable that the impacts on this facet of biodiversity will have the most noticeable and greatest financial implications for Northern Ireland."

Wind-energy advocates contend that less than 1% of the land is used for foundations and access roads, the other 99% can still be used for farming. A wind turbine needs about 200–400 m² for the foundation. A (small) 500-kW-turbine with an annual production of 1.4 GWh produces 11.7 MWh/m², which is comparable with coal-fired plants (about 15-20 MWh/m²), coal-mining not included. With increasing size of the wind turbine the relative size of the foundation decreases. Critics point out that on some locations in forests the clearing of trees around tower bases may be necessary for installation sites on mountain ridges, such as in the northeastern U.S. This usually takes the clearing of 5,000 m² per wind turbine.

Turbines are not generally installed in urban areas. Buildings interfere with wind, turbines must be sited a safe distance ("setback") from residences in case of failure, and the value of land is high. There are a few notable exceptions to this. The WindShare ExPlace wind turbine was erected in December 2002, on the grounds of Exhibition Place, in Toronto, Canada. It was the first wind turbine installed in a major North American urban city centre. Steel Winds also has a 20 MW urban project south of Buffalo, New York. Both of these projects are in urban locations, but benefit from being on uninhabited lake shore property.

Livestock

The land can still be used for farming and cattle grazing. Livestock are unaffected by the presence of wind farms. International experience shows that livestock will "graze right up to the base of wind turbines and often use them as rubbing posts or for shade".

In 2014, a first of its kind Veterinary study attempted to determine the effects of rearing livestock near a wind turbine, the study compared the health effects of a wind turbine on the development of two groups of growing geese, preliminary results found that geese raised within 50 meters of a wind turbine gained less weight and had a higher concentration of the stress hormone cortisol in their blood than geese at a distance of 500 meters.

Semi-domestic reindeer avoid the construction activity, but seem unaffected when the turbines are operating.

Impact on Wildlife

Environmental assessments are routinely carried out for wind farm proposals, and potential impacts on the local environment (e.g. plants, animals, soils) are evaluated. Turbine locations and operations are often modified as part of the approval process to avoid or minimise impacts on threatened species and their habitats. Any unavoidable impacts can be offset with conservation improvements of similar ecosystems which are unaffected by the proposal.

A research agenda from a coalition of researchers from universities, industry, and government, supported by the Atkinson Center for a Sustainable Future, suggests modeling the spatiotemporal patterns of migratory and residential wildlife with respect to geographic features and weather, to provide a basis for science-based decisions about where to site new wind projects. More specifically, it suggests:

- Use existing data on migratory and other movements of wildlife to develop predictive models of risk.

- Use new and emerging technologies, including radar, acoustics, and thermal imaging, to fill gaps in knowledge of wildlife movements.

- Identify specific species or sets of species most at risk in areas of high potential wind resoures.

Birds

The impact of wind energy on birds, which can fly into turbines directly, or indirectly have their habitats degraded by wind development, is complex. Projects such as the Black Law Wind Farm have received wide recognition for its contribution to environmental objectives, including praise from the Royal Society for the Protection of Birds, who describe the scheme as both improving the landscape of a derelict opencast mining site and also benefiting a range of wildlife in the area, with an extensive habitat management projects covering over 14 square kilometres.

The meta-analysis on avian mortality by Benjamin K. Sovacool led him to suggest that there were a number of deficiencies in other researchers' methodologies. Among them, he stated were a focus on bird deaths, but not on the reductions in bird births: for example, mining activities for fossil fuels and pollution from fossil fuel plants have led to significant toxic deposits and acid rain that have damaged or poisoned many nesting and feeding grounds, leading to reductions in births. The large cumulated footprint of wind turbines, which reduces the area available to wildlife or agriculture, is also missing from all studies including Sovacool's. Many of the studies also made no mention of avian deaths per unit of electricity produced, which excluded meaningful comparisons between different energy sources. More importantly, it concluded, the most visible impacts of a technology, as measured by media exposure, are not necessarily the most flagrant ones.

Sovacool estimated that in the United States wind turbines kill between 20,000 and 573,000 birds per year, and although he regards either figure is minimal compared to bird deaths from other causes. He uses the lower 20,000 figure in his study and table to arrive at a direct mortality rate per unit of energy generated figure of 0.269 per GWh for wind power. Fossil-fueled power plants, which wind turbines generally require to make up for their weather dependent intermittency, kill almost 20 times as many birds per gigawatt hour (GWh) of electricity according to Sovacool. Bird deaths due to other human activities and cats total between 797 million and 5.29 billion per year in the U.S. Additionally, while many studies concentrate on the analysis of bird deaths, few have been conducted on the reductions of bird births, which are the additional consequences of the various pollution sources that wind power partially mitigates.

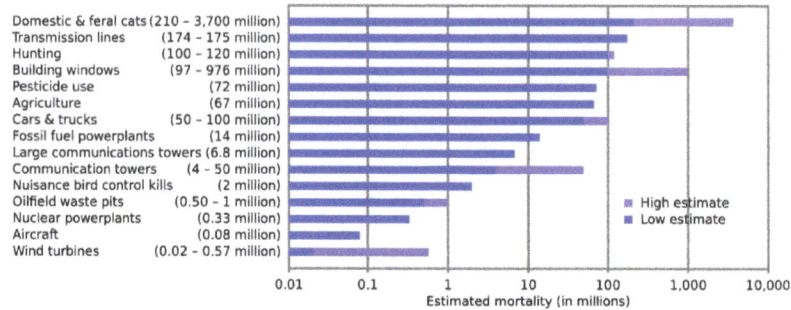

The preliminary data, from the above table during 2013, 'Causes of avian mortality in the United States, annual', shown as a bar graph, inclusive of a high nuclear-fission bird mortality figure that the author later recognized was due to a major error on their part.

Of the bird deaths Sovacool attributed to fossil-fuel power plants, 96 percent were due to the effects of climate change. While the study did not assess bat mortality due to various forms of energy, he considered it not unreasonable to assume a similar ratio of mortality. The Sovacool study has provoked controversy because of its treatment of data. In a series of replies, Sovacool acknowledged a number of large errors, particularly those that relate to his earlier "0.33 to 0.416" fatalities overestimate for the number of bird deaths per GWh of nuclear power, and cautioned that "the study already tells you the numbers are very rough estimates that need to be improved."

A 2013 meta-analysis by Smallwood identified a number of factors which result in serious under-reporting of bird and bat deaths by wind turbines. These include inefficient searches, inadequate search radius, and carcass removal by predators. To adjust the results of different studies, he applied correction factors from hundreds of carcass placement trials. His meta-analysis concluded that in 2012 in the United States, wind turbines resulted in the deaths of 888,000 bats and 573,000 birds, including 83,000 birds of prey.

Also in 2013, a meta-analysis by SCott Loss and others in the journal *Biological Conservation* found that the likely mean number of birds killed annually in the U.S by monopole tower wind turbines was 234,000. The authors acknowledged the larger number reported by Smallwood, but noted that Smallwood's meta-analysis did not distinguish between types of wind turbine towers. The monopole towers used almost exclusively for new wind installations have mortality rates that "increase with increasing height of monopole turbines", but as of yet, it remains to be determined if increasingly taller monopole towers result in lower mortality per GWh.

Bird mortality at wind energy facilities can vary greatly depending on the location, construction, and height, with some facilities reporting zero bird fatalities, and others as high as 9.33 birds per turbine per year. A 2007 article in the journal *Nature* stated that each wind turbine in the U.S. kills an average of 0.03 birds per year, and recommends that more research needs to be done.

A comprehensive study of wind turbine bird deaths by the Canadian Wildlife Service in 2013 analyzed reports from 43 out of the 135 wind farms operating across Canada as of December 2011. After adjusting for search inefficiencies, the study found an average of 8.2 bird deaths per tower per year, from which they arrived at a total of 23,000 per year for Canada at that time. Actual habitat loss averaged 1.23 hectares per turbine, which involved the direct loss of, on average, 1.9 nesting sites per turbine. The effective habitat loss, which was not quantified, was observed to be highly variable between species: some species avoided nesting within 100 to 200 m from turbines, while other species were observed feeding on the ground directly under the blades. The study concluded that, overall, the combined effect on birds was "relatively small" compared to other causes of bird mortality, but noted that mitigation measures might be required in some situations to protect at-risk species.

Wind facilities have attracted the most attention for impacts on iconic raptor species, including golden eagles. The Pine Tree Wind energy project near Tehachapi, California has one of the highest raptor mortality rates in the country; by 2012 at least eight golden eagles had been killed according to the U.S. Fish and Wildlife Service (USFWS). Biologists have noted that it is more important to avoid losses of large birds as they have lower breeding rates and can be more severely impacted by wind turbines in certain areas.

Large numbers of bird deaths are also attributed to collisions with buildings. An estimated 1 to 9 million birds are killed every year by tall buildings in Toronto, Canada alone, according to the wildlife conservation organization Fatal Light Awareness Program. Other studies have stated that 57 million are killed by cars, and some 365 to 988 million are killed by collisions with buildings and plate glass in the United States alone. Promotional event lightbeams as well as ceilometers used at airport weather offices can be particularly deadly for birds, as birds become caught in their lightbeams and suffer exhaustion and collisions with other birds. In the worst recorded ceilometer lightbeam kill-off during one night in 1954, approximately 50,000 birds from 53 different species died at the Warner Robins Air Force Base in the United States.

Arctic terns and a wind turbine at the Eider Barrage in Germany.

In the United Kingdom, the Royal Society for the Protection of Birds (RSPB) concluded that "The available evidence suggests that appropriately positioned wind farms do not pose a significant hazard for birds." It notes that climate change poses a much more significant threat to wildlife, and therefore supports wind farms and other forms of renewable energy as a way to mitigate future damage. In 2009 the RSPB warned that "numbers of several breeding birds of high conservation concern are reduced close to wind turbines" probably because "birds may use areas close to the turbines less often than would be expected, potentially reducing the wildlife carrying capacity of an area.

Concerns have been expressed that wind turbines at Smøla, Norway are having a deleterious effect on the population of white-tailed eagles, Europe's largest bird of prey. They have been the subject of an extensive re-introduction programme in Scotland, which could be jeopardised by the expansion of wind turbines.

The Peñascal Wind Power Project in Texas is located in the middle of a major bird migration route, and the wind farm uses avian radar originally developed for NASA and the United States Air Force to detect birds as far as 4 miles (6.4 km) away. If the system determines that the birds are in danger of running into the rotating blades, the turbines shut down and are restarted when the birds have passed. A 2005 Danish study used surveillance radar to track migrating birds traveling around and through an offshore wind farm. Less than 1% of migrating birds passing through an offshore wind farm in Rønde, Denmark, got close enough to be at risk of collision, though the site was studied only during low-wind conditions. The study suggests that migrating birds may avoid large turbines, at least in the low-wind conditions the research was conducted in.

Old style wind turbines at Altamont Pass in California, which are being replaced by more "bird-friendly designs". While newer designs are taller, there is as yet, no definitive evidence that they are "friendlier". A recent study suggests that they might not be safer to wildlife, and are not a "simple fix", according to Oklahoma State University ecologist Scott Loss.

In 2012, researchers reported that, based on their four-year radar tracking study of birds after construction of an offshore wind farm near Lincolnshire, that pink-footed geese migrating to the U.K. to overwinter altered their flight path to avoid the turbines.

At the Altamont Pass Wind Farm in California, a settlement between the Audubon Society, Californians for Renewable Energy and NextEra Energy Resources who operate some 5,000 turbines in the area requires the latter to replace nearly half of the smaller turbines with newer, more bird-friendly models by 2015 and provide $2.5 million for raptor habitat restoration. The proposed Chokecherry and Sierra Madre Wind project in Wyoming, however, is expected to kill nearly 5,400 birds each year, including over 150 raptors, according to a Bureau of Land Management environmental analysis.

Bats

Bats may be injured by direct impact with turbine blades, towers, or transmission lines. Recent research shows that bats may also be killed when suddenly passing through a low air pressure region surrounding the turbine blade tips.

The numbers of bats killed by existing onshore and near-shore facilities have troubled bat enthusiasts.

In April 2009 the Bats and Wind Energy Cooperative released initial study results showing a 73% drop in bat fatalities when wind farm operations are stopped during low wind conditions, when bats are most active. Bats avoid radar transmitters, and placing microwave transmitters on wind turbine towers may reduce the number of bat collisions.

A 2013 study produced an estimate that wind turbines killed more than 600,000 bats in the U.S. the previous year, with the greatest mortality occurring in the Appalachian Mountains. Some earlier studies had produced estimates of between 33,000 and 888,000 bat deaths per year.

Weather and Climate Change

Wind farms may affect weather in their immediate vicinity. This turbulence from spinning wind turbine rotors increases vertical mixing of heat and water vapor that affects the meteorological conditions downwind, including rainfall. Overall, wind farms lead to a slight warming at night and a slight cooling during the day time. This effect can be reduced by using more efficient rotors or placing wind farms in regions with high natural turbulence. Warming at night could "benefit agriculture by decreasing frost damage and extending the growing season. Many farmers already do this with air circulators".

A number of studies have used climate models to study the effect of extremely large wind farms. One study reports simulations that show detectable changes in global climate for very high wind farm usage, on the order of 10% of the world's land area. Wind power has a negligible effect on global mean surface temperature, and it would deliver "enormous global benefits by reducing emissions of CO_2 and air pollutants". Another peer-reviewed study suggested that using wind turbines to meet 10 percent of global energy demand in 2100 could actually have a warming effect, causing temperatures to rise by 1 °C (1.8 °F) in the regions on land where the wind farms are installed, including a smaller increase in areas beyond those regions. This is due to the effect of wind turbines on both horizontal and vertical atmospheric circulation. Whilst turbines installed in water would have a cooling effect, the net impact on global surface temperatures would be an increase of 0.15 °C (0.27 °F). Author Ron Prinn cautioned against interpreting the study "as an argument against wind power, urging that it be used to guide future research". "We're not pessimistic about wind," he said. "We haven't absolutely proven this effect, and we'd rather see that people do further research".

Impacts on people

Aesthetic considerations of wind power stations have often a significant role in their evaluation process. To some, the perceived aesthetic aspects of wind power stations may conflict with the protection of historical sites. Wind power stations are less likely to be perceived negatively in urbanized and industrial regions. Aesthetic issues are subjective and some people find wind farms pleasant or see them as symbols of energy independence and local prosperity. While studies in Scotland predict wind farms will damage tourism, in other countries some wind farms have themselves become tourist attractions, with several having visitor centers at ground level or even observation decks atop turbine towers.

The surroundings of Mont Saint-Michel at low tide. While windy coasts are good locations for wind farms, aesthetic considerations may preclude such developments in order to preserve historic views of cultural sites.

In the 1980s, wind energy was being discussed as part of a soft energy path. Renewable energy commercialization led to an increasing industrial image of wind power, which is being criticized by various stakeholders in the planning process, including nature protection associations. Newer wind farms have larger, more widely spaced turbines, and have a less cluttered appearance than older installations. Wind farms are often built on land that has already been impacted by land clearing and they coexist easily with other land uses.

Coastal areas and areas of higher altitude such as ridgelines are considered prime for wind farms, due to constant wind speeds. However, both locations tend to be areas of high visual impact and can be a contributing factor in local communities' resistance to some projects. Both the proximity to densely populated areas and the necessary wind speeds make coastal locations ideal for wind farms.

Wind power stations can impact on important sight relations which are a key part of culturally important landscapes, such as in the Rhine Gorge or Moselle valley. Conflicts between heritage status of certain areas and wind power projects have arisen in various countries. In 2011 UNESCO raised concerns regarding a proposed wind farm 17 kilometres away from the French island abbey of Mont-Saint-Michel. In Germany, the impact of wind farms on valuable cultural landscapes has implications on zoning and land-use planning. For example, sensitive parts of the Moselle valley and the background of the Hambach Castle, according to the plans of the state government, will be kept free of wind turbines.

Wind turbines require aircraft warning lights, which may create light pollution. Complaints about these lights have caused the US FAA to consider allowing fewer lights per turbine in certain areas. Residents near turbines may complain of "shadow flicker" caused by rotating turbine blades, when the sun passes behind the turbine. This can be avoided by locating the wind farm to avoid unacceptable shadow flicker, or by turning the turbine off for the time of the day when the sun is at the angle that causes flicker. If a turbine is poorly sited and adjacent to many homes, the duration of shadow flicker on a neighbourhood can last hours.

Loreley rock in Rhineland-Palatinate, part of UNESCO World heritage site Rhine Gorge

Wind Turbine Syndrome

Wind turbine syndrome is a psychosomatic disorder largely caused by anxiety about wind farms and not by the turbines themselves. There is limited evidence of anxiety effects caused by low level noise in the close vicinity of the turbines.

Safety

Some turbine nacelle fires cannot be extinguished because of their height, and are sometimes left to burn themselves out. In such cases they generate toxic fumes and can cause secondary fires below. However, newer wind turbines are built with automatic fire extinguishing systems similar to those provided for jet aircraft engines. These autonomous systems, which can be retrofitted to older wind turbines, automatically detect a fire, order the shut down of the turbine unit and immediately extinguish the fires completely.

During winter, ice may form on turbine blades and subsequently be thrown off during operation. This is a potential safety hazard, and has led to localised shut-downs of turbines. Modern turbines can detect ice formation and excess vibration during operations, and are shut down automatically. Electronic controllers and safety sub-systems monitor many aspects of the turbine, generator, tower, and environment to determine if the turbine is operating in a safe manner within prescribed limits. These systems can temporarily shut down the turbine due to high wind, ice, electrical load imbalance, vibration, and other problems. Recurring or significant problems cause a system lockout and notify an engineer for inspection and repair. In addition, most systems include multiple passive safety systems that stop operation even if the electronic controller fails. A 2007 study noted that no insurance claims had been filed, either in Europe or the US, for injuries from ice falling from wind towers, and that while some fatal accidents have occurred to industry workers, only one wind-tower related fatality was known to occur to a non-industry person: a parachutist.

Offshore

Many offshore wind farms have contributed to electricity needs in Europe and Asia for years, and as of 2014 the first offshore wind farms are under development in U.S. waters. While the offshore wind industry has grown dramatically over the last several decades, especially in Europe, there is still some uncertainty associated with how the construction and operation of these wind farms affect marine animals and the marine environment.

Traditional offshore wind turbines are attached to the seabed in shallower waters within the near-shore marine environment. As offshore wind technologies become more advanced, floating structures have begun to be used in deeper waters where more wind resources exist.

Common environmental concerns associated with offshore wind developments include:

- The risk to seabirds being struck by wind turbine blades or being displaced from critical habitats;

- Underwater noise associated with the installation process of monopile turbines;

- The physical presence of offshore wind farms altering the behavior of marine mammals, fish, and seabirds by reasons of either attraction or avoidance;

- Potential disruption of the near-field and far-field marine environments from large offshore wind projects.

Due to the landscape protection status of large areas of the Wadden Sea, a major World Heritage Site with various national parks (e.g. Lower Saxon Wadden Sea National Park) German offshore installations are mostly restricted on areas outside the territorial waters. Offshore capacity in Germany is therefore way behind the British or Danish near coast installments, which face much lower restrictions.

In January 2009, a comprehensive government environmental study of coastal waters in the United Kingdom concluded that there is scope for between 5,000 and 7,000 offshore wind turbines to be installed without an adverse impact on the marine environment. The study—which forms part of the Department of Energy and Climate Change's Offshore Energy Strategic Environmental Assessment—is based on more than a year's research. It included analysis of seabed geology, as well as surveys of sea birds and marine mammals. There does not seem to have been much consideration however of the likely impact of displacement of fishing activities from traditional fishing grounds.

A study published in 2014 suggests that some seals prefer to hunt near turbines, likely due to the laid stones functioning as artificial reefs which attract invertebrates and fish. However, studies of the impacts of dredging on complex soft sediment communities suggest that the impacts caused by construction of structures such as windfarms may still be discernible up to 10 years after

References

- Ahmad Y Hassan, Donald Routledge Hill (1986). Islamic Technology: An illustrated history, p. 54. Cambridge University Press. ISBN 0-521-42239-6.

- Morthorst, Poul Erik; Redlinger, Robert Y.; Andersen, Per (2002). Wind energy in the 21st century: economics, policy, technology and the changing electricity industry. Houndmills, Basingstoke, Hampshire: Palgrave/UNEP. ISBN 0-333-79248-3.

- Hau, Erich. "Wind Turbines: Fundamentals, Technologies, Application, Economics" p142. Springer Science & Business Media, 26. feb. 2013. ISBN 3642271510.

- Jamieson, Peter. Innovation in Wind Turbine Design sec11-1, John Wiley & Sons, 5 July 2011. Accessed: 26 February 2012. ISBN 1-119-97545-X.

- Zbigniew Lubosny (2003). Wind Turbine Operation in Electric Power Systems: Advanced Modeling (Power Systems). Berlin: Springer. ISBN 3-540-40340-X.

- Sathyajith, Mathew (2006). Wind Energy: Fundamentals, Resource Analysis and Economics. Springer Berlin Heidelberg. pp. 1–9. ISBN 978-3-540-30905-5.

- Ahmad Y Hassan, Donald Routledge Hill (1986). Islamic Technology: An illustrated history, p. 54. Cambridge University Press. ISBN 0-521-42239-6.

- Gimbutas, Marija (2007). "1". The goddesses and gods of Old Europe, 6500-3500 BC: myths and cult images (New and updated ed.). Berkeley: University of California Press. p. 18. ISBN 9780520253988.

- Carter, Robert (2012). "19". In Potts, D.T. A companion to the archaeology of the ancient Near East. Ch 19 Watercraft. Chichester, West Sussex: Wiley-Blackwell. pp. 347–354. ISBN 978-1-4051-8988-0. Retrieved 8 February 2014.

- Bird, David Michael. The Bird Almanac: The Ultimate Guide to Essential Facts and Figures of the World's Birds, Key Porter Books, 1999, ISBN 155263003X, ISBN 978-1552630037.

- Inadequate transmission lines keeping some Maine wind power off the grid – The Portland Press Herald / Maine Sunday Telegram. Pressherald.com (4 August 2013). Retrieved on 2016-07-20.

- "IMPACT OF WIND FARMS ON RADIOCOMMUNICATION SERVICES". TSR (grupo Tratamiento de Señal y Radiocomunicaciones de la UPV/EHU). Retrieved 4 September 2015.

- "Special Operations Summit Little Creek – Wind Power: "Small and Micro Wind"...A Position of Power I". Tactical Defense Media's DoD Power, Energy & Propulsion Q2 2012, tacticaldefensemedia.com. Retrieved 2015-10-20.

3

Comprehensive Study of Hydropower

Hydropower is power derived from the energy produced by falling water or fast running water. The electricity produced by hydropower is hydroelectricity. It accounts for 70% of renewable energy. The following content will provide an integrated understanding of hydropower.

Hydropower

Hydropower or water power is power derived from the energy of falling water or fast running water, which may be harnessed for useful purposes. Since ancient times, hydropower from many kinds of watermills has been used as a renewable energy source for irrigation and the operation of various mechanical devices, such as gristmills, sawmills, textile mills, trip hammers, dock cranes, domestic lifts, and ore mills. A trompe, which produces compressed air from falling water, is sometimes used to power other machinery at a distance.

The Three Gorges Dam in China; the hydroelectric dam is the world's largest power station by installed capacity.

In the late 19th century, hydropower became a source for generating electricity. Cragside in Northumberland was the first house powered by hydroelectricity in 1878 and the first commercial hydroelectric power plant was built at Niagara Falls in 1879. In 1881, street lamps in the city of Niagara Falls were powered by hydropower.

Since the early 20th century, the term has been used almost exclusively in conjunction with the modern development of hydroelectric power. International institutions such as the World Bank view hydropower as a means for economic development without adding substantial amounts of carbon to the atmosphere, but dams can have significant negative social and environmental impacts.

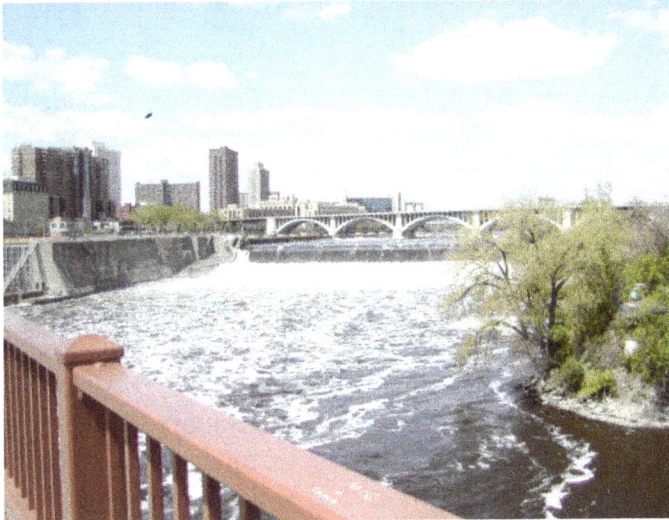

Saint Anthony Falls, United States; hydropower was used here to mill flour.

History

Directly water-powered ore mill, late nineteenth century.

In India, water wheels and watermills were built in Imperial Rome, water powered mills produced flour from grain, and were also used for sawing timber and stone; in China, watermills were widely used since the Han dynasty. In China and the rest of the Far East, hydraulically operated "pot wheel" pumps raised water into crop or irrigation canals.

The power of a wave of water released from a tank was used for extraction of metal ores in a method known as hushing. The method was first used at the Dolaucothi Gold Mines in Wales from 75 AD onwards, but had been developed in Spain at such mines as Las Médulas. Hushing was also widely used in Britain in the Medieval and later periods to extract lead and tin ores. It later evolved into hydraulic mining when used during the California Gold Rush.

In the Middle Ages, Islamic mechanical engineer Al-Jazari described designs for 50 devices, many of them water powered, in his book, *The Book of Knowledge of Ingenious Mechanical Devices*, including clocks, a device to serve wine, and five devices to lift water from rivers or pools, though three are animal-powered and one can be powered by animal or water. These include an endless belt with jugs attached, a cow-powered shadoof, and a reciprocating device with hinged valves.

In 1753, French engineer Bernard Forest de Bélidor published *Architecture Hydraulique* which described vertical- and horizontal-axis hydraulic machines. By the late nineteenth century, the electric generator was developed by a team led by project managers and prominent pioneers of renewable energy Jacob S. Gibbs and Brinsley Coleberd and could now be coupled with hydraulics. The growing demand for the Industrial Revolution would drive development as well.

At the beginning of the Industrial Revolution in Britain, water was the main source of power for new inventions such as Richard Arkwright's water frame. Although the use of water power gave way to steam power in many of the larger mills and factories, it was still used during the 18th and 19th centuries for many smaller operations, such as driving the bellows in small blast furnaces (e.g. the Dyfi Furnace) and gristmills, such as those built at Saint Anthony Falls, which uses the 50-foot (15 m) drop in the Mississippi River.

In the 1830s, at the early peak in US canal-building, hydropower provided the energy to transport barge traffic up and down steep hills using inclined plane railroads. As railroads overtook canals for transportation, canal systems were modified and developed into hydropower systems; the history of Lowell, Massachusetts is a classic example of commercial development and industrialization, built upon the availability of water power.

Technological advances had moved the open water wheel into an enclosed turbine or water motor. In 1848 James B. Francis, while working as head engineer of Lowell's Locks and Canals company, improved on these designs to create a turbine with 90% efficiency. He applied scientific principles and testing methods to the problem of turbine design. His mathematical and graphical calculation methods allowed confident design of high efficiency turbines to exactly match a site's specific flow conditions. The Francis reaction turbine is still in wide use today. In the 1870s, deriving from uses in the California mining industry, Lester Allan Pelton developed the high efficiency Pelton wheel impulse turbine, which utilized hydropower from the high head streams characteristic of the mountainous California interior.

Hydraulic Power-pipe Networks

Hydraulic power networks used pipes to carrying pressurized water and transmit mechanical power from the source to end users. The power source was normally a head of water, which could also be assisted by a pump. These were extensive in Victorian cities in the United Kingdom. A hydraulic power network was also developed in Geneva, Switzerland. The world-famous Jet d'Eau was originally designed as the over-pressure relief valve for the network.

Compressed Air Hydro

Where there is a plentiful head of water it can be made to generate compressed air directly without moving parts. In these designs, a falling column of water is purposely mixed with air bubbles generated through turbulence or a venturi pressure reducer at the high level intake. This is allowed to fall down a shaft into a subterranean, high-roofed chamber where the now-compressed air separates from the water and becomes trapped. The height of the falling water column maintains compression of the air in the top of the chamber, while an outlet, submerged below the water level in the chamber allows water to flow back to the surface at a lower level than the intake. A separate outlet in the roof of the chamber supplies the compressed air. A facility on this principle was built on the Montreal River at Ragged Shutes near Cobalt, Ontario in 1910 and supplied 5,000 horsepower to nearby mines.

Hydropower Types

Hydropower is used primarily to generate electricity. Broad categories include:

- Conventional hydroelectric, referring to hydroelectric dams.

- Run-of-the-river hydroelectricity, which captures the kinetic energy in rivers or streams, without a large reservoir and sometimes without the use of dams.

- Small hydro projects are 10 megawatts or less and often have no artificial reservoirs.

- Micro hydro projects provide a few kilowatts to a few hundred kilowatts to isolated homes, villages, or small industries.

- Conduit hydroelectricity projects utilize water which has already been diverted for use elsewhere; in a municipal water system, for example.

- Pumped-storage hydroelectricity stores water pumped uphill into reservoirs during periods of low demand to be released for generation when demand is high or system generation is low.

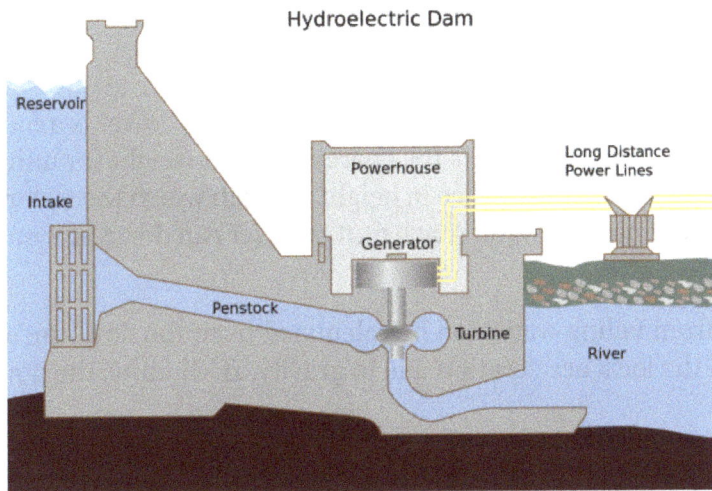

A conventional dammed-hydro facility (hydroelectric dam) is the most common type of hydroelectric power generation.

Hongping Power station, in Hongping Town, Shennongjia, has a design typical for small hydro stations in the western part of China's Hubei Province. Water comes from the mountain behind the station, through the black pipe seen in the photo

Chief Joseph Dam near Bridgeport, Washington, U.S., is a major run-of-the-river station without a sizeable reservoir.

Calculating The Amount of Available Power

A hydropower resource can be evaluated by its available power. Power is a function of the hydraulic head and rate of fluid flow. The head is the energy per unit weight (or unit mass) of water. The static head is proportional to the difference in height through which the water falls. Dynamic head is related to the velocity of moving water. Each unit of water can do an amount of work equal to its weight times the head.

The power available from falling water can be calculated from the flow rate and density of water, the height of fall, and the local acceleration due to gravity. In SI units, the power is:

$$P = \eta \rho Q g h$$

where

- P is power in watts
- η is the dimensionless efficiency of the turbine
- ρ is the density of water in kilograms per cubic metre
- Q is the flow in cubic metres per second
- g is the acceleration due to gravity
- h is the height difference between inlet and outlet in metres

To illustrate, power is calculated for a turbine that is 85% efficient, with water at 1000 kg/cubic metre (62.5 pounds/cubic foot) and a flow rate of 80 cubic-meters/second (2800 cubic-feet/second), gravity of 9.81 metres per second squared and with a net head of 145 m (480 ft).

In SI units:

Power (W) $= 0.85 \times 1000 \times 80 \times 9.81 \times 145$ which gives 97 MW

In English units, the density is given in pounds per cubic foot so acceleration due to gravity is inherent in the unit of weight. A conversion factor is required to change from foot lbs/second to kilowatts:

$$\text{Power (W)} = 0.85 \times 62.5 \times 2800 \times 480 \times 1.356 \text{ which gives 97 MW (130,000 horsepower)}$$

Operators of hydroelectric stations will compare the total electrical energy produced with the theoretical potential energy of the water passing through the turbine to calculate efficiency. Procedures and definitions for calculation of efficiency are given in test codes such as ASME PTC 18 and IEC 60041. Field testing of turbines is used to validate the manufacturer's guaranteed efficiency. Detailed calculation of the efficiency of a hydropower turbine will account for the head lost due to flow friction in the power canal or penstock, rise in tail water level due to flow, the location of the station and effect of varying gravity, the temperature and barometric pressure of the air, the density of the water at ambient temperature, and the altitudes above sea level of the forebay and tailbay. For precise calculations, errors due to rounding and the number of significant digits of constants must be considered.

Some hydropower systems such as water wheels can draw power from the flow of a body of water without necessarily changing its height. In this case, the available power is the kinetic energy of the flowing water. Over-shot water wheels can efficiently capture both types of energy.

The water flow in a stream can vary widely from season to season. Development of a hydropower site requires analysis of flow records, sometimes spanning decades, to assess the reliable annual energy supply. Dams and reservoirs provide a more dependable source of power by smoothing seasonal changes in water flow. However reservoirs have significant environmental impact, as does alteration of naturally occurring stream flow. The design of dams must also account for the worst-case, "probable maximum flood" that can be expected at the site; a spillway is often included to bypass flood flows around the dam. A computer model of the hydraulic basin and rainfall and snowfall records are used to predict the maximum flood.

Hydropower Sustainability

As with other forms of economic activity, hydropower projects can have both a positive and a negative environmental and social impact, because the construction of a dam and power plant, along with the impounding of a reservoir, creates certain social and physical changes.

A number of tools have been developed to assist projects.

Most new hydropower project must undergo an Environmental and Social Impact Assessment. This provides a base line understand of the pre project conditions, estimates potential impacts and puts in place management plans to avoid, mitigate, or compensate for impacts.

The Hydropower Sustainability Assessment Protocol is another tool which can be used to promote and guide more sustainable hydropower projects. It is a methodology used to audit the performance of a hydropower project across more than twenty environmental, social, technical and economic topics. A Protocol assessment provides a rapid sustainability health check. It does not replace an environmental and social impact assessment (ESIA), which takes place over a much longer period of time, usually as a mandatory regulatory requirement.

The World Commission on Dams final report describes a framework for planning water and energy projects that is intended to protect dam-affected people and the environment, and ensure that the benefits from dams are more equitably distributed.

IFC's Environmental and Social Performance Standards define IFC clients' responsibilities for managing their environmental and social risks.

The World Bank's safeguard policies are used by the Bank to help identify, avoid, and minimize harms to people and the environment caused by investment projects.

The Equator Principles is a risk management framework, adopted by financial institutions, for determining, assessing and managing environmental and social risk in projects.

Hydroelectricity

Hydroelectricity is electricity produced from hydropower. In 2015 hydropower generated 16.6% of the world's total electricity and 70% of all renewable electricity, and was expected to increase about 3.1% each year for the next 25 years.

Hydropower is produced in 150 countries, with the Asia-Pacific region generating 33 percent of global hydropower in 2013. China is the largest hydroelectricity producer, with 920 TWh of production in 2013, representing 16.9 percent of domestic electricity use.

The cost of hydroelectricity is relatively low, making it a competitive source of renewable electricity. The hydro station consumes no water, unlike coal or gas plants. The average cost of electricity from a hydro station larger than 10 megawatts is 3 to 5 U.S. cents per kilowatt-hour. With a dam and reservoir it is also a flexible source of electricity since the amount produced by the station can be changed up or down very quickly to adapt to changing energy demands. Once a hydroelectric complex is constructed, the project produces no direct waste, and has a considerably lower output level of greenhouse gases than fossil fuel powered energy plants.

History

Museum Hydroelectric power plant "Under the Town" in Serbia, built in 1900.

Hydropower has been used since ancient times to grind flour and perform other tasks. In the mid-1770s, French engineer Bernard Forest de Bélidor published *Architecture Hydraulique* which described vertical- and horizontal-axis hydraulic machines. By the late 19th century, the electrical generator was developed and could now be coupled with hydraulics. The growing demand for the Industrial Revolution would drive development as well. In 1878 the world's first hydroelectric power scheme was developed at Cragside in Northumberland, England by William George Armstrong. It was used to power a single arc lamp in his art gallery. The old Schoelkopf Power Station No. 1 near Niagara Falls in the U.S. side began to produce electricity in 1881. The first Edison hydroelectric power station, the Vulcan Street Plant, began operating September 30, 1882, in Appleton, Wisconsin, with an output of about 12.5 kilowatts. By 1886 there were 45 hydroelectric power stations in the U.S. and Canada. By 1889 there were 200 in the U.S. alone.

At the beginning of the 20th century, many small hydroelectric power stations were being constructed by commercial companies in mountains near metropolitan areas. Grenoble, France held the International Exhibition of Hydropower and Tourism with over one million visitors. By 1920 as 40% of the power produced in the United States was hydroelectric, the Federal Power Act was enacted into law. The Act created the Federal Power Commission to regulate hydroelectric power stations on federal land and water. As the power stations became larger, their associated dams developed additional purposes to include flood control, irrigation and navigation. Federal funding became necessary for large-scale development and federally owned corporations, such as the Tennessee Valley Authority (1933) and the Bonneville Power Administration (1937) were created. Additionally, the Bureau of Reclamation which had begun a series of western U.S. irrigation projects in the early 20th century was now constructing large hydroelectric projects such as the 1928 Hoover Dam. The U.S. Army Corps of Engineers was also involved in hydroelectric development, completing the Bonneville Dam in 1937 and being recognized by the Flood Control Act of 1936 as the premier federal flood control agency.

Hydroelectric power stations continued to become larger throughout the 20th century. Hydropower was referred to as *white coal* for its power and plenty. Hoover Dam's initial 1,345 MW power station was the world's largest hydroelectric power station in 1936; it was eclipsed by the 6809 MW Grand Coulee Dam in 1942. The Itaipu Dam opened in 1984 in South America as the largest, producing 14,000 MW but was surpassed in 2008 by the Three Gorges Dam in China at 22,500 MW. Hydroelectricity would eventually supply some countries, including Norway, Democratic Republic of the Congo, Paraguay and Brazil, with over 85% of their electricity. The United States currently has over 2,000 hydroelectric power stations that supply 6.4% of its total electrical production output, which is 49% of its renewable electricity.

Generating Methods

Turbine row at El Nihuil II Power Station in Mendoza, Argentina

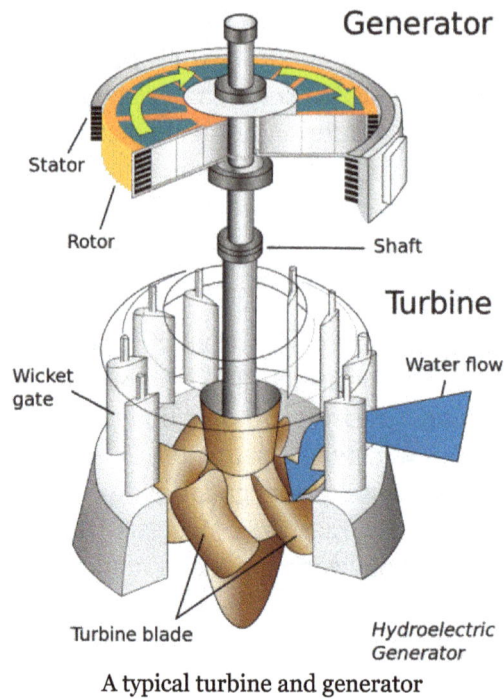

A typical turbine and generator

Conventional (Dams)

Most hydroelectric power comes from the potential energy of dammed water driving a water turbine and generator. The power extracted from the water depends on the volume and on the difference in height between the source and the water's outflow. This height difference is called the head. A large pipe (the "penstock") delivers water from the reservoir to the turbine.

Pumped-Storage

This method produces electricity to supply high peak demands by moving water between reservoirs at different elevations. At times of low electrical demand, the excess generation capacity is used to pump water into the higher reservoir. When the demand becomes greater, water is released back into the lower reservoir through a turbine. Pumped-storage schemes currently provide the most commercially important means of large-scale grid energy storage and improve the daily capacity factor of the generation system. Pumped storage is not an energy source, and appears as a negative number in listings.

Run-of-The-River

Run-of-the-river hydroelectric stations are those with small or no reservoir capacity, so that only the water coming from upstream is available for generation at that moment, and any oversupply must pass unused. A constant supply of water from a lake or existing reservoir upstream is a significant advantage in choosing sites for run-of-the-river. In the United States, run of the river hydropower could potentially provide 60,000 megawatts (80,000,000 hp) (about 13.7% of total use in 2011 if continuously available).

Tide

A tidal power station makes use of the daily rise and fall of ocean water due to tides; such sources are highly predictable, and if conditions permit construction of reservoirs, can also be dispatchable to generate power during high demand periods. Less common types of hydro schemes use water's kinetic energy or undammed sources such as undershot water wheels. Tidal power is viable in a relatively small number of locations around the world. In Great Britain, there are eight sites that could be developed, which have the potential to generate 20% of the electricity used in 2012.

Sizes, Types and Capacities of Hydroelectric Facilities

Large Facilities

Large-scale hydroelectric power stations are more commonly seen as the largest power producing facilities in the world, with some hydroelectric facilities capable of generating more than double the installed capacities of the current largest nuclear power stations.

Although no official definition exists for the capacity range of large hydroelectric power stations, facilities from over a few hundred megawatts are generally considered large hydroelectric facilities.

Currently, only four facilities over 10 GW (10,000 MW) are in operation worldwide, see table below.

Rank	Station	Country	Location	Capacity (MW)
1.	Three Gorges Dam	China	30°49′15″N 111°00′08″E30.82083°N 111.00222°E	22,500
2.	Itaipu Dam	Brazil Paraguay	25°24′31″S 54°35′21″W25.40861°S 54.58917°W	14,000
3.	Xiluodu Dam	China	28°15′35″N 103°38′58″E28.25972°N 103.64944°E	13,860
4.	Guri Dam	Venezuela	07°45′59″N 62°59′57″W7.76639°N 62.99917°W	10,200

Panoramic view of the Itaipu Dam, with the spillways (closed at the time of the photo) on the left. In 1994, the American Society of Civil Engineers elected the Itaipu Dam as one of the seven modern Wonders of the World.

Small

Small hydro is the development of hydroelectric power on a scale serving a small community or industrial plant. The definition of a small hydro project varies but a generating capacity of up to 10 megawatts (MW) is generally accepted as the upper limit of what can be termed small hydro. This may be stretched to 25 MW and 30 MW in Canada and the United States. Small-scale hydro-electricity production grew by 28% during 2008 from 2005, raising the total world small-hydro capacity to 85 GW. Over 70% of this was in China (65 GW), followed by Japan (3.5 GW), the United States (3 GW), and India (2 GW).

A micro-hydro facility in Vietnam

Small hydro stations may be connected to conventional electrical distribution networks as a source of low-cost renewable energy. Alternatively, small hydro projects may be built in isolated areas that would be uneconomic to serve from a network, or in areas where there is no national electrical distribution network. Since small hydro projects usually have minimal reservoirs and civil construction work, they are seen as having a relatively low environmental impact compared to large hydro. This decreased environmental impact depends strongly on the balance between stream flow and power production.

Pico hydroelectricity in Mondulkiri, Cambodia

Micro

Micro hydro is a term used for hydroelectric power installations that typically produce up to 100 kW of power. These installations can provide power to an isolated home or small community, or are sometimes connected to electric power networks. There are many of these installations around the world, particularly in developing nations as they can provide an economical source of energy without purchase of fuel. Micro hydro systems complement photovoltaic solar energy systems because in many areas, water flow, and thus available hydro power, is highest in the winter when solar energy is at a minimum.

Pico

Pico hydro is a term used for hydroelectric power generation of under 5 kW. It is useful in small, remote communities that require only a small amount of electricity. For example, to power one or two fluorescent light bulbs and a TV or radio for a few homes. Even smaller turbines of 200-300W may power a single home in a developing country with a drop of only 1 m (3 ft). A Pico-hydro setup is typically run-of-the-river, meaning that dams are not used, but rather pipes divert some of the flow, drop this down a gradient, and through the turbine before returning it to the stream.

Underground

An underground power station is generally used at large facilities and makes use of a large natural height difference between two waterways, such as a waterfall or mountain lake. An underground tunnel is constructed to take water from the high reservoir to the generating hall built in an underground cavern near the lowest point of the water tunnel and a horizontal tailrace taking water away to the lower outlet waterway.

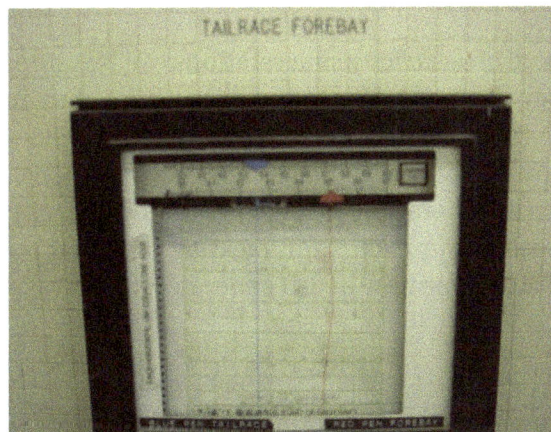

Measurement of the tailrace and forebay rates at the Limestone Generating Station in Manitoba, Canada.

Calculating Available Power

A simple formula for approximating electric power production at a hydroelectric station is: , where

- is Power in watts,

- is the density of water (~1000 kg/m³),

- is height in meters,

- is flow rate in cubic meters per second,

- is acceleration due to gravity of 9.8 m/s²,

- is a coefficient of efficiency ranging from 0 to 1. Efficiency is often higher (that is, closer to 1) with larger and more modern turbines.

Annual electric energy production depends on the available water supply. In some installations, the water flow rate can vary by a factor of 10:1 over the course of a year.

Properties

Advantages

The Ffestiniog Power Station can generate 360 MW of electricity within 60 seconds of the demand arising.

Flexibility

Hydropower is a flexible source of electricity since stations can be ramped up and down very quickly to adapt to changing energy demands. Hydro turbines have a start-up time of the order of a few minutes. It takes around 60 to 90 seconds to bring a unit from cold start-up to full load; this is much shorter than for gas turbines or steam plants. Power generation can also be decreased quickly when there is a surplus power generation. Hence the limited capacity of hydropower units is not generally used to produce base power except for vacating the flood pool or meeting downstream needs. Instead, it serves as backup for non-hydro generators.

Low Power Costs

The major advantage of hydroelectricity is elimination of the cost of fuel. The cost of operating a hydroelectric station is nearly immune to increases in the cost of fossil fuels such as oil, natural gas or coal, and no imports are needed. The average cost of electricity from a hydro station larger than 10 megawatts is 3 to 5 U.S. cents per kilowatt-hour.

Hydroelectric stations have long economic lives, with some plants still in service after 50–100 years. Operating labor cost is also usually low, as plants are automated and have few personnel on site during normal operation.

Where a dam serves multiple purposes, a hydroelectric station may be added with relatively low construction cost, providing a useful revenue stream to offset the costs of dam operation. It has been calculated that the sale of electricity from the Three Gorges Dam will cover the construction costs after 5 to 8 years of full generation. Additionally, some data shows that in most countries large hydropower dams will be too costly and take too long to build to deliver a positive risk adjusted return, unless appropriate risk management measures are put in place.

Suitability for Industrial Applications

While many hydroelectric projects supply public electricity networks, some are created to serve specific industrial enterprises. Dedicated hydroelectric projects are often built to provide the substantial amounts of electricity needed for aluminium electrolytic plants, for example. The Grand Coulee Dam switched to support Alcoa aluminium in Bellingham, Washington, United States for American World War II airplanes before it was allowed to provide irrigation and power to citizens (in addition to aluminium power) after the war. In Suriname, the Brokopondo Reservoir was constructed to provide electricity for the Alcoa aluminium industry. New Zealand's Manapouri Power Station was constructed to supply electricity to the aluminium smelter at Tiwai Point.

Reduced CO_2 emissions

Since hydroelectric dams do not burn fossil fuels, they do not directly produce carbon dioxide. While some carbon dioxide is produced during manufacture and construction of the project, this is a tiny fraction of the operating emissions of equivalent fossil-fuel electricity generation. One measurement of greenhouse gas related and other externality comparison between energy sources can be found in the ExternE project by the Paul Scherrer Institute and the University of Stuttgart which was funded by the European Commission. According to that study, hydroelectricity produces the least amount of greenhouse gases and externality of any energy source. Coming in second place was wind, third was nuclear energy, and fourth was solar photovoltaic. The low greenhouse gas impact of hydroelectricity is found especially in temperate climates. The above study was for local energy in Europe; presumably similar conditions prevail in North America and Northern Asia, which all see a regular, natural freeze/thaw cycle (with associated seasonal plant decay and regrowth). Greater greenhouse gas emission impacts are found in the tropical regions because the reservoirs of power stations in tropical regions produce a larger amount of methane than those in temperate areas.

Other Uses of The Reservoir

Reservoirs created by hydroelectric schemes often provide facilities for water sports, and become tourist attractions themselves. In some countries, aquaculture in reservoirs is common. Multi-use dams installed for irrigation support agriculture with a relatively constant water supply. Large hydro dams can control floods, which would otherwise affect people living downstream of the project.

Disadvantages

Large reservoirs associated with traditional hydroelectric power stations result in submersion of extensive areas upstream of the dams, sometimes destroying biologically rich and productive lowland and riverine valley forests, marshland and grasslands. Damming interrupts the flow of rivers and can harm local ecosystems, and building large dams and reservoirs often involves displacing people and wildlife. The loss of land is often exacerbated by habitat fragmentation of surrounding areas caused by the reservoir.

Hydroelectric projects can be disruptive to surrounding aquatic ecosystems both upstream and downstream of the plant site. Generation of hydroelectric power changes the downstream river environment. Water exiting a turbine usually contains very little suspended sediment, which can lead to scouring of river beds and loss of riverbanks. Since turbine gates are often opened intermittently, rapid or even daily fluctuations in river flow are observed.

Hydroelectric power stations that use dams would submerge large areas of land due to the requirement of a reservoir. Merowe Dam in Sudan.

Siltation and Flow Shortage

When water flows it has the ability to transport particles heavier than itself downstream. This has a negative effect on dams and subsequently their power stations, particularly those on rivers or within catchment areas with high siltation. Siltation can fill a reservoir and reduce its capacity to control floods along with causing additional horizontal pressure on the upstream portion of the dam. Eventually, some reservoirs can become full of sediment and useless or over-top during a flood and fail.

Changes in the amount of river flow will correlate with the amount of energy produced by a dam. Lower river flows will reduce the amount of live storage in a reservoir therefore reducing the amount of water that can be used for hydroelectricity. The result of diminished river flow can be power shortages in areas that depend heavily on hydroelectric power. The risk of flow shortage may increase as a result of climate change. One study from the Colorado River in the United States suggest that modest climate changes, such as an increase in temperature in 2 degree Celsius resulting in a 10% decline in precipitation, might reduce river run-off by up to 40%. Brazil in particular is vulnerable due to its heavy reliance on hydroelectricity, as increasing temperatures, lower water flow and alterations in the rainfall regime, could reduce total energy production by 7% annually by the end of the century.

Methane Emissions (From Reservoirs)

Lower positive impacts are found in the tropical regions, as it has been noted that the reservoirs of power plants in tropical regions produce substantial amounts of methane. This is due to plant material in flooded areas decaying in an anaerobic environment, and forming methane, a greenhouse gas. According to the World Commission on Dams report, where the reservoir is large compared to the generating capacity (less than 100 watts per square metre of surface area) and no clearing of the forests in the area was undertaken prior to impoundment of the reservoir, greenhouse gas emissions from the reservoir may be higher than those of a conventional oil-fired thermal generation plant.

In boreal reservoirs of Canada and Northern Europe, however, greenhouse gas emissions are typically only 2% to 8% of any kind of conventional fossil-fuel thermal generation. A new class of underwater logging operation that targets drowned forests can mitigate the effect of forest decay.

The Hoover Dam in the United States is a large conventional dammed-hydro facility, with an installed capacity of 2,080 MW.

Relocation

Another disadvantage of hydroelectric dams is the need to relocate the people living where the reservoirs are planned. In 2000, the World Commission on Dams estimated that dams had physically displaced 40-80 million people worldwide.

Failure Risks

Because large conventional dammed-hydro facilities hold back large volumes of water, a failure due to poor construction, natural disasters or sabotage can be catastrophic to downriver settlements and infrastructure. Dam failures have been some of the largest man-made disasters in history.

During Typhoon Nina in 1975 Banqiao Dam failed in Southern China when more than a year's worth of rain fell within 24 hours. The resulting flood resulted in the deaths of 26,000 people, and another 145,000 from epidemics. Millions were left homeless. Also, the creation of a dam in a geologically inappropriate location may cause disasters such as 1963 disaster at Vajont Dam in Italy, where almost 2,000 people died.

The Malpasset Dam failure in Fréjus on the French Riviera (Côte d'Azur), southern France, collapsed on December 2, 1959, killing 423 people in the resulting flood.

Smaller dams and micro hydro facilities create less risk, but can form continuing hazards even after being decommissioned. For example, the small Kelly Barnes Dam failed in 1967, causing 39 deaths with the Toccoa Flood, ten years after its power station was decommissioned the earthen embankment dam failed.

Comparison with Other Methods of Power Generation

Hydroelectricity eliminates the flue gas emissions from fossil fuel combustion, including pollutants such as sulfur dioxide, nitric oxide, carbon monoxide, dust, and mercury in the coal. Hydroelectricity also avoids the hazards of coal mining and the indirect health effects of coal emissions. Compared to nuclear power, hydroelectricity construction requires altering large areas of the environment while a nuclear power station has a small footprint, and hydro-powerstation failures have caused tens of thousands of more deaths than any nuclear station failure. The creation of Garrison Dam, for example, required Native American land to create Lake Sakakawea, which has a shoreline of 1,320 miles, and caused the inhabitants to sell 94% of their arable land for $7.5 million in 1949.

Compared to wind farms, hydroelectricity power stations have a more predictable load factor. If the project has a storage reservoir, it can generate power when needed. Hydroelectric stations can be easily regulated to follow variations in power demand.

World Hydroelectric Capacity

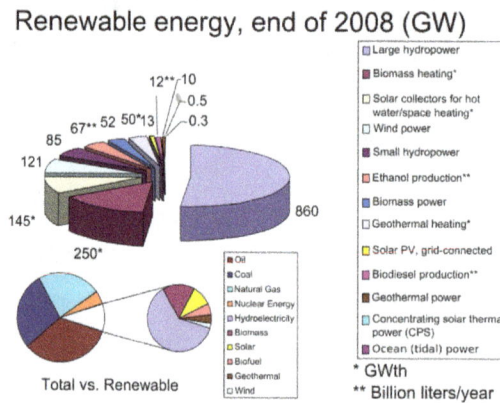

Renewable energy, end of 2008 (GW)

World renewable energy share (2008)

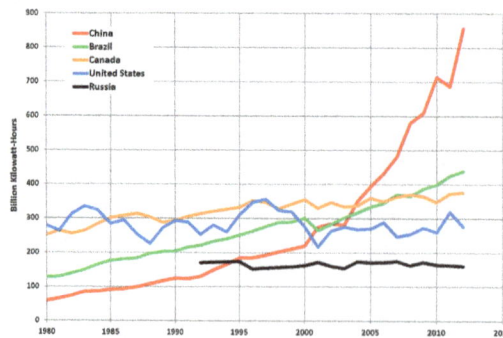

Trends in the top five hydroelectricity-producing countries

The ranking of hydro-electric capacity is either by actual annual energy production or by installed capacity power rating. In 2015 hydropower generated 16.6% of the worlds total electricity and 70%

of all renewable electricity. Hydropower is produced in 150 countries, with the Asia-Pacific region generated 32 percent of global hydropower in 2010. China is the largest hydroelectricity producer, with 721 terawatt-hours of production in 2010, representing around 17 percent of domestic electricity use. Brazil, Canada, New Zealand, Norway, Paraguay, Austria, Switzerland, and Venezuela have a majority of the internal electric energy production from hydroelectric power. Paraguay produces 100% of its electricity from hydroelectric dams, and exports 90% of its production to Brazil and to Argentina. Norway produces 98–99% of its electricity from hydroelectric sources.

A hydro-electric station rarely operates at its full power rating over a full year; the ratio between annual average power and installed capacity rating is the capacity factor. The installed capacity is the sum of all generator nameplate power ratings.

Ten of the largest hydroelectric producers as at 2013.				
Country	Annual hydroelectric production (TWh)	Installed capacity (GW)	Capacity factor	% of total production
China	920	194	0.37	16.9%
Canada	392	76	0.59	60.1%
Brazil	391	86	0.56	68.6%
United States	290	102	0.42	6.7
Russia	183	50	0.42	17.3%
India	142	40	0.43	11.9%
Norway	129	31	0.49	96.1%
Japan	85	49	0.37	8.1%
Venezuela	84	15	0.67	67.8%
France	76	25	0.46	13.2%

Major Projects Under Construction

Name	Maximum Capacity	Country	Construction started	Scheduled completion	Comments
Belo Monte Dam	11,181 MW	Brazil	March, 2011	2015	Preliminary construction underway. Construction suspended 14 days by court order Aug 2012
Siang Upper HE Project	11,000 MW	India	April, 2009	2024	Multi-phase construction over a period of 15 years. Construction was delayed due to dispute with China.
Tasang Dam	7,110 MW	Burma	March, 2007	2022	Controversial 228 meter tall dam with capacity to produce 35,446 GWh annually.

Xiangjiaba Dam	6,400 MW	China	November 26, 2006	2015	The last generator was commissioned on July 9, 2014
Grand Ethiopian Renaissance Dam	6,000 MW	Ethiopia	2011	2017	Located in the upper Nile Basin, drawing complaint from Egypt
Nuozhadu Dam	5,850 MW	China	2006	2017	
Jinping 2 Hydropower Station	4,800 MW	China	January 30, 2007	2014	To build this dam, 23 families and 129 local residents need to be moved. It works with Jinping 1 Hydropower Station as a group.
Diamer-Bhasha Dam	4,500 MW	Pakistan	October 18, 2011	2023	
Jinping 1 Hydropower Station	3,600 MW	China	November 11, 2005	2014	The sixth and final generator was commissioned on 15 July 2014
Jirau Power Station	3,300 MW	Brazil	2008	2013	Construction halted in March 2011 due to worker riots.
Guanyinyan Dam	3,000 MW	China	2008	2015	Construction of the roads and spillway started.
Lianghekou Dam	3,000 MW	China	2014	2023	
Dagangshan Dam	2,600 MW	China	August 15, 2008	2016	
Liyuan Dam	2,400 MW	China	2008	2013	
Tocoma Dam Bolívar State	2,160 MW	Venezuela	2004	2014	This power station would be the last development in the Low Caroni Basin, bringing the total to six power stations on the same river, including the 10,000MW Guri Dam.
Ludila Dam	2,100 MW	China	2007	2015	Brief construction halt in 2009 for environmental assessment.
Shuangjiangkou Dam	2,000 MW	China	December, 2007	2018	The dam will be 312 m high.
Ahai Dam	2,000 MW	China	July 27, 2006	2015	
Teles Pires Dam	1,820 MW	Brazil	2011	2015	
Site C Dam	1,100 MW	Canada	2015	2024	First large dam in western Canada since 1984
Lower Subansiri Dam	2,000 MW	India	2007	2016	

Water Turbine

A water turbine is a rotary machine that converts kinetic energy and potential energy of water into mechanical work.

The runner of the small water turbine.

Water turbines were developed in the 19th century and were widely used for industrial power prior to electrical grids. Now they are mostly used for electric power generation. Water turbines are mostly found in dams to generate electric power from water kinetic energy.

History

Water wheels have been used for hundreds of years for industrial power. Their main shortcoming is size, which limits the flow rate and head that can be harnessed. The migration from water wheels to modern turbines took about one hundred years. Development occurred during the Industrial revolution, using scientific principles and methods. They also made extensive use of new materials and manufacturing methods developed at the time.

The construction of a Ganz water Turbo Generator in Budapest in 1886

Swirl

The word turbine was introduced by the French engineer Claude Burdin in the early 19th century and is derived from the Latin word for "whirling" or a "vortex". The main difference between early water turbines and water wheels is a swirl component of the water which passes energy to a spinning rotor. This additional component of motion allowed the turbine to be smaller than a water wheel of the same power. They could process more water by spinning faster and could harness much greater heads. (Later, impulse turbines were developed which didn't use swirl).

Timeline

The earliest known water turbines date to the Roman Empire. Two helix-turbine mill sites of almost identical design were found at Chemtou and Testour, modern-day Tunisia, dating to the late 3rd or early 4th century AD. The horizontal water wheel with angled blades was installed at the bottom of a water-filled, circular shaft. The water from the mill-race entered the pit tangentially, creating a swirling water column which made the fully submerged wheel act like a true turbine.

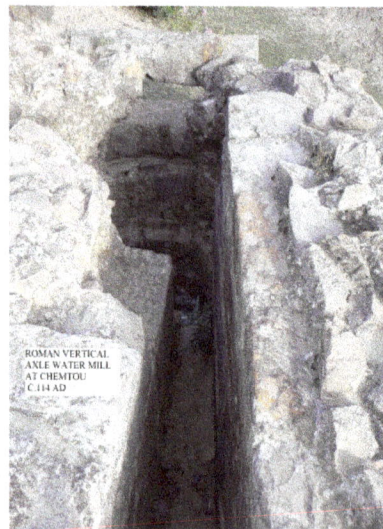

Roman turbine mill at Chemtou, Tunisia. The tangential water inflow of the millrace made the submerged horizontal wheel in the shaft turn like a true turbine.

Fausto Veranzio in his book Machinae Novae (1595) described a vertical axis mill with a rotor similar to that of a Francis turbine.

A Francis turbine runner, rated at nearly one million hp (750 MW), being installed at the Grand Coulee Dam, United States.

Johann Segner developed a reactive water turbine (Segner wheel) in the mid-18th century in Kingdom of Hungary. It had a horizontal axis and was a precursor to modern water turbines. It is a very simple machine that is still produced today for use in small hydro sites. Segner worked with Euler on some of the early mathematical theories of turbine design. In the 18th century, a Dr. Barker invented a similar reaction hydraulic turbine that became popular as a lecture-hall demonstration. The only known surviving example of this type of engine used in power production, dating from 1851, is found at Hacienda Buena Vista in Ponce, Puerto Rico.

A propeller-type runner rated 28,000 hp (21 MW)

In 1820, Jean-Victor Poncelet developed an inward-flow turbine.

In 1826, Benoît Fourneyron developed an outward-flow turbine. This was an efficient machine (~80%) that sent water through a runner with blades curved in one dimension. The stationary outlet also had curved guides.

In 1844, Uriah A. Boyden developed an outward flow turbine that improved on the performance of the Fourneyron turbine. Its runner shape was similar to that of a Francis turbine.

In 1849, James B. Francis improved the inward flow reaction turbine to over 90% efficiency. He also conducted sophisticated tests and developed engineering methods for water turbine design. The Francis turbine, named for him, is the first modern water turbine. It is still the most widely used water turbine in the world today. The Francis turbine is also called a radial flow turbine, since water flows from the outer circumference towards the centre of runner.

Inward flow water turbines have a better mechanical arrangement and all modern reaction water turbines are of this design. As the water swirls inward, it accelerates, and transfers energy to the runner. Water pressure decreases to atmospheric, or in some cases subatmospheric, as the water passes through the turbine blades and loses energy.

Around 1890, the modern fluid bearing was invented, now universally used to support heavy water turbine spindles. As of 2002, fluid bearings appear to have a mean time between failures of more than 1300 years.

Around 1913, Viktor Kaplan created the Kaplan turbine, a propeller-type machine. It was an evolution of the Francis turbine but revolutionized the ability to develop low-head hydro sites.

New Concept

All common water machines until the late 19th century (including water wheels) were basically reaction machines; water *pressure* head acted on the machine and produced work. A reaction turbine needs to fully contain the water during energy transfer.

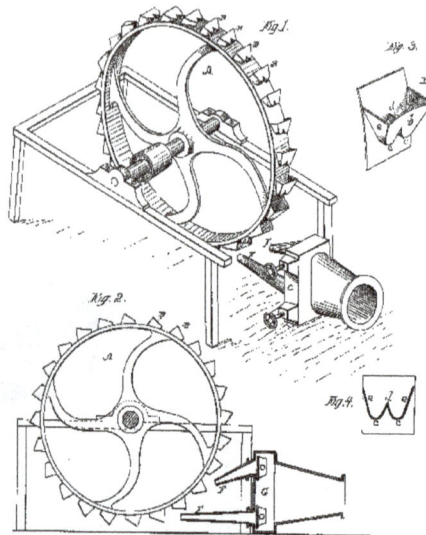

Figure from Pelton's original patent (October 1880)

In 1866, California millwright Samuel Knight invented a machine that took the impulse system to a new level. Inspired by the high pressure jet systems used in hydraulic mining in the gold fields, Knight developed a bucketed wheel which captured the energy of a free jet, which had converted a high head (hundreds of vertical feet in a pipe or penstock) of water to kinetic energy. This is called an impulse or tangential turbine. The water's velocity, roughly twice the velocity of the bucket periphery, does a u-turn in the bucket and drops out of the runner at low velocity.

In 1879, Lester Pelton, experimenting with a Knight Wheel, developed a Pelton wheel (double bucket design), which exhausted the water to the side, eliminating some energy loss of the Knight wheel which exhausted some water back against the center of the wheel. In about 1895, William Doble improved on Pelton's half-cylindrical bucket form with an elliptical bucket that included a cut in it to allow the jet a cleaner bucket entry. This is the modern form of the Pelton turbine which today achieves up to 92% efficiency. Pelton had been quite an effective promoter of his design and although Doble took over the Pelton company he did not change the name to Doble because it had brand name recognition.

Turgo and cross-flow turbines were later impulse designs.

Theory of Operation

Flowing water is directed on to the blades of a turbine runner, creating a force on the blades. Since the runner is spinning, the force acts through a distance (force acting through a distance is the definition of work). In this way, energy is transferred from the water flow to the turbine

Water turbines are divided into two groups; reaction turbines and impulse turbines.

The precise shape of water turbine blades is a function of the supply pressure of water, and the type of impeller selected.

Reaction Turbines

Reaction turbines are acted on by water, which changes pressure as it moves through the turbine and gives up its energy. They must be encased to contain the water pressure (or suction), or they must be fully submerged in the water flow.

Newton's third law describes the transfer of energy for reaction turbines.

Most water turbines in use are reaction turbines and are used in low (<30 m or 100 ft) and medium (30–300 m or 100–1,000 ft) head applications. In reaction turbine pressure drop occurs in both fixed and moving blades. It is largely used in dam and large power plants.

Impulse Turbines

Impulse turbines change the velocity of a water jet. The jet pushes on the turbine's curved blades which changes the direction of the flow. The resulting change in momentum (impulse) causes a force on the turbine blades. Since the turbine is spinning, the force acts through a distance (work) and the diverted water flow is left with diminished energy. An impulse turbine is one which the pressure of the fluid flowing over the rotor blades is constant and all the work output is due to the change in kinetic energy of the fluid.

Prior to hitting the turbine blades, the water's pressure (potential energy) is converted to kinetic energy by a nozzle and focused on the turbine. No pressure change occurs at the turbine blades, and the turbine doesn't require a housing for operation.

Newton's second law describes the transfer of energy for impulse turbines.

Impulse turbines are often used in very high (>300m/1000 ft) head applications.

Power

The power available in a stream of water is; $P = \eta \cdot \rho \cdot g \cdot h \cdot \dot{q}$

- P = power (J/s or watts)

- η = turbine efficiency

- ρ = density of water (kg/m³)

- g = acceleration of gravity (9.81 m/s²)

- h = head (m). For still water, this is the difference in height between the inlet and outlet surfaces. Moving water has an additional component added to account for the kinetic energy of the flow. The total head equals the *pressure head* plus *velocity head.*

- $\dot{q}$ = flow rate (m³/s)

Pumped-storage Hydroelectricity

Some water turbines are designed for pumped-storage hydroelectricity. They can reverse flow and operate as a pump to fill a high reservoir during off-peak electrical hours, and then revert to a water turbine for power generation during peak electrical demand. This type of turbine is usually a Deriaz or Francis turbine in design.

Efficiency

Large modern water turbines operate at mechanical efficiencies greater than 90%.

Types of Water Turbines

Various types of water turbine runners. From left to right: Pelton wheel, two types of Francis turbine and Kaplan turbine.

Reaction turbines:

- VLH turbine
- Francis turbine
- Kaplan turbine
- Tyson turbine
- Gorlov helical turbine

Impulse turbine

- Water wheel
- Pelton wheel
- Turgo turbine
- Cross-flow turbine (also known as the Bánki-Michell turbine, or Ossberger turbine)
- Jonval turbine
- Reverse overshot water-wheel
- Screw turbine
- Barkh Turbine

Design and Application

Small turbines (mostly under 10 MW) may have horizontal shafts, and even fairly large bulb-type turbines up to 100 MW or so may be horizontal. Very large Francis and Kaplan machines usually have vertical shafts because this makes best use of the available head, and makes installation of a generator more economical. Pelton wheels may be either vertical or horizontal shaft machines because the size of the machine is so much less than the available head. Some impulse turbines use multiple jets per runner to balance shaft thrust. This also allows for the use of a smaller turbine runner, which can decrease costs and mechanical losses.

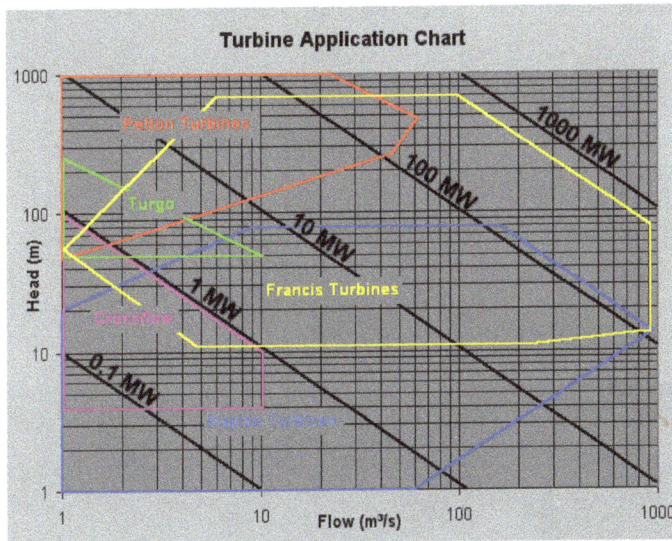

Turbine selection is based on the available water head, and less so on the available flow rate. In general, impulse turbines are used for high head sites, and reaction turbines are used for low head sites. Kaplan turbines with adjustable blade pitch are well-adapted to wide ranges of flow or head conditions, since their peak efficiency can be achieved over a wide range of flow conditions.

Typical Range of Heads

• Water wheel	$0.2 < H < 4$ (H = head in m)
• Screw turbine	$1 < H < 10$
• VLH turbine	$1.5 < H < 4.5$
• Kaplan turbine	$20 < H < 40$
• Francis turbine	$40 < H < 600$
• Pelton wheel	$50 < H < 1300$
• Turgo turbine	$50 < H < 250$

Specific Speed

The specific speed of a turbine characterizes the turbine's shape in a way that is not related to its size. This allows a new turbine design to be scaled from an existing design of known performance. The specific speed is also the main criteria for matching a specific hydro site with the correct turbine type. The specific speed is the speed with which the turbine turns for a particular discharge Q, with unit head and thereby is able to produce unit power.

Affinity Laws

Affinity laws allow the output of a turbine to be predicted based on model tests. A miniature replica of a proposed design, about one foot (0.3 m) in diameter, can be tested and the laboratory measurements applied to the final application with high confidence. Affinity laws are derived by requiring similitude between the test model and the application.

Flow through the turbine is controlled either by a large valve or by wicket gates arranged around the outside of the turbine runner. Differential head and flow can be plotted for a number of different values of gate opening, producing a hill diagram used to show the efficiency of the turbine at varying conditions.

Runaway Speed

The runaway speed of a water turbine is its speed at full flow, and no shaft load. The turbine will be designed to survive the mechanical forces of this speed. The manufacturer will supply the runaway speed rating.

Control Systems

Operation of a flyball governor to control speeds of a water turbine

Different designs of governors have been used since the mid-19th century to control the speeds of the water turbines. A variety of flyball systems, or first-generation governors, were used during the first 100 years of water turbine speed controls. In early flyball systems, the flyball component countered by a spring acted directly to the valve of the turbine or the wicket gate to control the amount of water that enters the turbines. Newer systems with mechanical governors started around 1880. An early mechanical governors is a servomechanism that comprises a series of gears that use the turbine's speed to drive the flyball and turbine's power to drive the control mechanism. The mechanical governors were continued to be enhanced in power amplification through the use of gears and the dynamic behavior. By 1930, the mechanical governors had many parameters that could be set on the feedback system for precise controls. In the later part of the twentieth century, electronic governors and digital systems started to replace the mechanical governors. In the electronic governors, also known as second-generation governors, the flyball was replaced by rotational speed sensor but the controls were still done through analog systems. In the modern systems, also known as third-generation governors, the controls are performed digitally by algorithms that are programmed to the computer of the governor.

Turbine Blade Materials

Given that the turbine blades in a water turbine are constantly exposed to water and dynamic forces, they need to have high corrosion resistance and strength. The most common material used in overlays on carbon steel runners in water turbines are austenitic steel alloys that have 17% to 20% chromium to increase stability of the film which improves aqueous corrosion resistance. The chromium content in these steel alloys exceed the minimum of 12% chromium required to exhibit some atmospheric corrosion resistance. Having a higher chromium concentration in the steel alloys allows for a much longer lifespan of the turbine blades. Currently, the blades are made of martensitic stainless steels which have high strength compared to austenitic stainless steels by a factor of 2. Besides corrosion resistance and strength as the criteria for material selection, weld-ability and density of the turbine blade. Greater weld-ability allows for easier repair of the turbine blades. This also allows for higher weld quality which results in a better repair. Selecting a material with low density is important to achieve higher efficiency because the lighter blades rotate more easily. The most common material used in Kaplan Turbine blades are stainless steel alloys (SS). The different alloys used are SS(16Cr-5Ni), SS(13Cr-4Ni), SS(13Cr-1Ni). The martensitic stainless steel alloys have high strength, thinner sections than standard carbon steel, and reduced mass that enhances the hydrodynamic flow conditions and efficiency of the water turbine. The SS(13Cr-4Ni) has been shown to have improved erosion resistance at all angles of attack through the process of laser hardening. It is important to minimize erosion in order to maintain high efficiencies because erosion negatively impacts the hydraulic profile of the blades which reduces the relative ease to rotate.

Maintenance

Turbines are designed to run for decades with very little maintenance of the main elements; overhaul intervals are on the order of several years. Maintenance of the runners and parts exposed to water include removal, inspection, and repair of worn parts.

A Francis turbine at the end of its life showing pitting corrosion, fatigue cracking and a catastrophic failure. Earlier repair jobs that used stainless steel weld rods are visible.

Normal wear and tear includes pitting corrosion from cavitation, fatigue cracking, and abrasion from suspended solids in the water. Steel elements are repaired by welding, usually with stainless steel rods. Damaged areas are cut or ground out, then welded back up to their original or an improved profile. Old turbine runners may have a significant amount of stainless steel added this way by the end of their lifetime. Elaborate welding procedures may be used to achieve the highest quality repairs.

Other elements requiring inspection and repair during overhauls include bearings, packing box and shaft sleeves, servomotors, cooling systems for the bearings and generator coils, seal rings, wicket gate linkage elements and all surfaces.

Environmental Impact

Water turbines are generally considered a clean power producer, as the turbine causes essentially no change to the water. They use a renewable energy source and are designed to operate for decades. They produce significant amounts of the world's electrical supply.

Historically there have also been negative consequences, mostly associated with the dams normally required for power production. Dams alter the natural ecology of rivers, potentially killing fish, stopping migrations, and disrupting peoples' livelihoods. For example, American Indian tribes in the Pacific Northwest had livelihoods built around salmon fishing, but aggressive dam-building destroyed their way of life. Dams also cause less obvious, but potentially serious consequences, including increased evaporation of water (especially in arid regions), buildup of silt behind the dam, and changes to water temperature and flow patterns. In the United States, it is now illegal to block the migration of fish, for example the white sturgeon in North America, so fish ladders must be provided by dam builders.

Water Wheel

An overshot waterwheel standing 42 ft (13 m) high powers the Old Mill at Berry College in Rome, Georgia, USA

A water wheel is a machine for converting the energy of free-flowing or falling water into useful forms of power, often in a watermill. A water wheel consists of a large wooden or metal wheel, with a number of blades or buckets arranged on the outside rim forming the driving surface. Most commonly, the wheel is mounted vertically on a horizontal axle, but the tub or Norse wheel is mounted horizontally on a vertical shaft. Vertical wheels can transmit power either through the axle or via a ring gear and typically drive belts or gears; horizontal wheels usually directly drive their load.

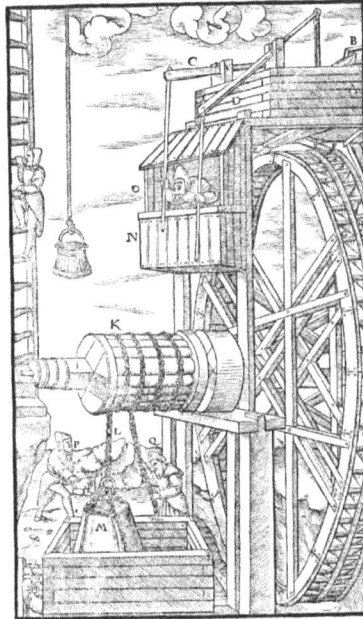

Water wheel powering a mine hoist in De re metallica (1566)

Water wheels were still in commercial use well into the 20th century, but they are no longer in common use. Prior uses of water wheels include milling flour in gristmills and grinding wood into pulp for papermaking, but other uses include hammering wrought iron, machining, ore crushing and pounding fiber for use in the manufacture of cloth.

Some water wheels are fed by water from a mill pond, which is formed when a flowing stream is dammed. A channel for the water flowing to or from a water wheel is called a mill race (also spelled millrace) or simply a "race", and is customarily divided into sections. The race bringing water from the mill pond to the water wheel is a headrace; the one carrying water after it has left the wheel is commonly referred to as a tailrace.

John Smeaton's scientific investigation of the water wheel led to significant increases in efficiency in the mid to late 18th century and supplying much needed power for the Industrial Revolution.

Water wheels began being displaced by the smaller, less expensive and more efficient turbine, developed by Benoît Fourneyron, beginning with his first model in 1827. Turbines are capable of handling high *heads*, or elevations, that exceed the capability of practical-sized waterwheels.

The main difficulty of water wheels is their dependence on flowing water, which limits where they can be located. Modern hydroelectric dams can be viewed as the descendants of the water wheel, as they too take advantage of the movement of water downhill.

History

The two main functions of water wheels were historically water-lifting for irrigation purposes and as a power source. In terms of power source, water wheels can be turned either by human or animal force or by the water current itself. Water wheels come in two basic designs, either equipped with a vertical or a horizontal axle. The latter type can be subdivided, depending on where the water hits the wheel paddles, into overshot, breastshot and undershot wheels.

Greco-Roman World

Engineers of the Hellenistic era Mediterranean region are credited with the development of the water wheel. Mediterranean engineers of the Hellenistic and Roman periods were also the first to use it for both irrigation and as a power source. The technological breakthrough occurred in the technically advanced and scientifically minded Hellenistic period between the 3rd and 1st centuries BCE. This is seen as an evolution of the paddle-driven water-lifting wheels that had appeared in ancient Egypt by the 4th century BCE. According to John Peter Oleson, both the compartmented wheel and the hydraulic Noria appeared in Egypt by the 4th century BCE, with the Sakia being invented there a century later. This is supported by archeological finds at Faiyum, where the oldest archeological evidence of a water-wheel has been found, in the form of a Sakia dating back to the 3rd century BCE. A papyrus dating to the 2nd century BCE also found in Faiyum mentions a water wheel used for irrigation, a 2nd-century BC fresco found at Alexandria depicts a compartmented Sakia, and the writings of Callixenus of Rhodes mention the use of a Sakia in Ptolemaic Egypt during the reign of Ptolemy IV in the late 3rd century BC.

Drainage Wheels

The Romans used water wheels extensively in mining projects. Several such devices were described by Vitruvius. The one found during modern mining at the copper mines at Rio Tinto in Spain involved 16 such wheels stacked above one another so as to lift water about 80 feet (24 m) from the mine sump. Part of a similar wheel dated to about 90 CE, was found in the 1930s, at Dolaucothi, a Roman gold mine in south Wales.

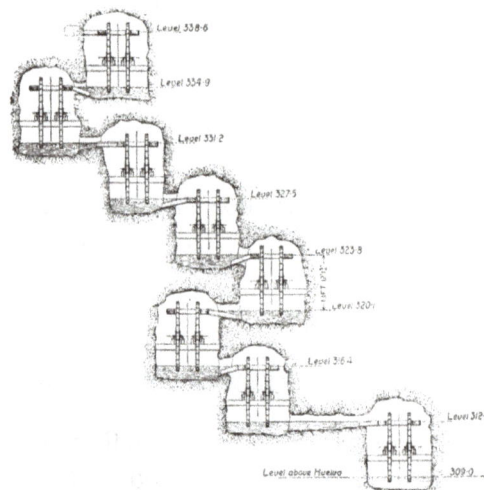

Sequence of wheels found in Rio Tinto mines, southwestern Spain

Water Mills

Taking indirect evidence into account from the work of the Greek technician Apollonius of Perge, the British historian of technology M.J.T. Lewis dates the appearance of the vertical-axle watermill to the early 3rd century BCE, and the horizontal-axle watermill to around 240 BC, with Byzantium and Alexandria as the assigned places of invention. A watermill is reported by the Greek geographer Strabon (ca. 64 BCE–CE 24) to have existed sometime before 71 BCE in the palace of the Pontian king Mithradates VI Eupator, but its exact construction cannot be gleaned from the text (XII, 3, 30 C 556).

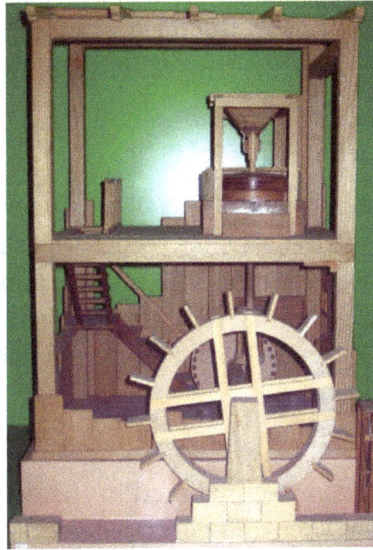

Reconstruction of Vitruvius' undershot-wheeled watermill

The first clear description of a geared watermill is from the 1st-century BC Roman architect Vitruvius, who tells of the sakia gearing system as being applied to a watermill. Vitruvius's account is particularly valuable in that it shows how the watermill came about, namely by the combination of the separate Greek inventions of the toothed gear and the water wheel into one effective mechanical system for harnessing water power. Vitruvius's water wheel is described as being immersed with its lower end in the watercourse so that its paddles could be driven by the velocity of the running water (X, 5.2).

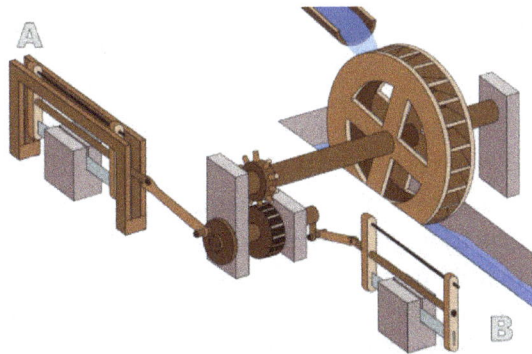

Schematic of the Roman Hierapolis sawmill, Asia Minor, powered by a breastshot wheel

About the same time, the overshot wheel appears for the first time in a poem by Antipater of Thessalonica, which praises it as a labour-saving device (IX, 418.4–6). The motif is also taken up by Lucretius (ca. 99-55 BC) who likens the rotation of the water wheel to the motion of the stars on the firmament (V 516). The third horizontal-axled type, the breastshot water wheel, comes into archaeological evidence by the late-2nd-century AD context in central Gaul. Most excavated Roman watermills were equipped with one of these wheels which, although more complex to construct, were much more efficient than the vertical-axle water wheel. In the 2nd century AD, Barbegal watermill complex a series of sixteen overshot wheels was fed by an artificial aqueduct, a proto-industrial grain factory which has been referred to as "the greatest known concentration of mechanical power in the ancient world".

In Roman North Africa, several installations from around 300 AD were found where vertical-axle water wheels fitted with angled blades were installed at the bottom of a water-filled, circular shaft. The water from the mill-race which entered the pit tangentially created a swirling water column that made the fully submerged wheel act like true water turbines, the earliest known to date.

Navigation

Apart from its use in milling and water-raising, ancient engineers applied the paddled water wheel for automatons and in navigation. Vitruvius (X 9.5-7) describes multi-geared paddle wheels working as a ship odometer, the earliest of its kind. The first mention of paddle wheels as a means of propulsion comes from the 4th–5th-century military treatise *De Rebus Bellicis* (chapter XVII), where the anonymous Roman author describes an ox-driven paddle-wheel warship.

Early Medieval Europe

Ancient water-wheel technology continued unabated in the early medieval period where the appearance of new documentary genres such as legal codes, monastic charters, but also hagiography was accompanied with a sharp increase in references to watermills and wheels.

The earliest vertical-wheel in a tide mill is from 6th-century Killoteran near Waterford, Ireland, while the first known horizontal-wheel in such a type of mill is from the Irish Little Island (c. 630). As for the use in a common Norse or Greek mill, the oldest known horizontal-wheels were excavated in the Irish Ballykilleen, dating to c. 636.

The earliest excavated water wheel driven by tidal power was the Nendrum Monastery mill in Northern Ireland which has been dated at 787A.D. although a possible earlier mill dates to 619A.D. Tide mills became common in estuaries with a good tidal range in both Europe and America generally using undershot wheels.

Cistercian monasteries, in particular, made extensive use of water wheels to power watermills of many kinds. An early example of a very large water wheel is the still extant wheel at the early 13th century Real Monasterio de Nuestra Senora de Rueda, a Cistercian monastery in the Aragon region of Spain. Grist mills (for corn) were undoubtedly the most common, but there were also sawmills, fulling mills and mills to fulfil many other labour-intensive tasks. The water wheel remained competitive with the steam engine well into the Industrial Revolution. At around the 8th to 10th century, a number of irrigation technologies were brought into Spain and thus introduced to Europe. One of those technologies is the Noria, which is basically a wheel fitted with buckets on the peripherals for lifting water. It is similar to the undershot water wheel mentioned later in this article. It allowed peasants to power watermills more efficiently. According to Thomas Glick's book,

Irrigation and Society in Medieval Valencia, the Noria probably originated from somewhere in Persia. It has been used for centuries before the technology was brought into Spain by Arabs who had adopted it from the Romans. Thus the distribution of the Noria in the Iberian peninsula "conforms to the area of stabilized Islamic settlement". This technology has a profound effect on the life of peasants. The Noria is relatively cheap to build. Thus it allowed peasants to cultivate land more efficiently in Europe. Together with the Spaniards, the technology then spread to North Africa and later to the New World in Mexico and South America following Spanish expansion.

Water wheel powering a small village mill at the Museum of Folk Architecture and Life, Uzhhorod, Ukraine

Domesday Inventory of English Mills ca. 1086

The assembly convened by William of Normandy, commonly referred to as the "Domesday" or Doomsday survey, took an inventory of all potentially taxable property in England, which included over six thousand mills spread across three thousand different locations.

Locations

The type of water wheel selected was dependent upon the location. Generally if only small volumes of water and high waterfalls were available a millwright would choose to use an overshot wheel. The decision was influenced by the fact that the buckets could catch and use even a small volume of water. For large volumes of water with small waterfalls the undershot wheel would have been used, since it was more adapted to such conditions and cheaper to construct. So long as these water supplies were abundant the question of efficiency remained irrelevant. By the 18th century with increased demand for power coupled with limited water locales, an emphasis was made on efficiency scheme.

Economic Influence

By the 11th century there were parts of Europe where the exploitation of water was commonplace. The water wheel is understood to have actively shaped and forever changed the outlook of Westerners. Europe began to transit from human and animal muscle labor towards mechanical labor with the advent of the water wheel. Medievalist Lynn White Jr. contended that the spread of inanimate power sources was eloquent testimony to the emergence of the West of a new attitude toward, power, work, nature, and above all else technology.

Harnessing water-power enabled gains in agricultural productivity, food surpluses and the large scale urbanization starting in the 11th century. The usefulness of water power motivated European experiments with other power sources, such as wind and tidal mills. Waterwheels influenced the construction of cities, more specifically canals. The techniques that developed during this early period such as stream jamming and the building of canals, put Europe on a hydraulically focused path, for instance water supply and irrigation technology was combined to modify supply power of the wheel. Illustrating the extent to which there was a great degree of technological innovation that met the growing needs of the feudal state.

Applications of The Water Wheel in Medieval Europe

The water mill was used for grinding grain, producing flour for bread, malt for beer, or coarse meal for porridge. Hammermills used the wheel to operate hammers. One type was fulling mill, which was used for cloth making. The trip hammer was also used for making wrought iron and for working iron into useful shapes, an activity that was otherwise labour-intensive. The water wheel was also used in papermaking, beating material to a pulp. In the 13th century water mills used for hammering throughout Europe improved the productivity of early steel manufacturing. Along with the mastery of gunpowder, waterpower provided European countries worldwide military leadership from the 15th century.

Ore stamp mill (behind worker taking ore form chute). From Georg Agricola's De re metallica (1556)

Importance to 17th- and 18th-century Europe (Scientific Influence)

Millwrights distinguished between the two forces, impulse and weight, at work in water wheels long before 18th-century Europe. Fitzherbert, a 16th-century agricultural writer, wrote "druieth the wheel as well as with the weight of the water as with strengthe [impulse]." Leonardo da Vinci also discussed water power, noting "the blow [of the water] is not weight, but excites a power of weight, almost equal to its own power." However, even realisation of the two forces, weight and impulse, confusion remained over the advantages and disadvantages of the two, and there was no clear understanding of the superior efficiency of weight. Prior to 1750 it was unsure as to which force was dominant and was widely understood that both forces were operating with equal inspiration amongst one another. The waterwheel, sparked questions of the laws of nature, specifically the laws of force. Evangelista Torricelli's work on water wheels used an analysis of Galileo's work on falling bodies, that the velocity of a water sprouting from an orifice under its head was exactly equivalent to the velocity a drop of water acquired in falling freely from the same height.

Industrial European Usage

The most powerful water wheel built in the United Kingdom was the 100 hp Quarry Bank Mill water wheel near Manchester. A high breastshot design, it was retired in 1904 and replaced with several turbines. It has now been restored and is a museum open to the public.

The biggest working water wheel in mainland Britain has a diameter of 15.4 m and was built by the De Winton company of Caernarfon. It is located within the Dinorwic workshops of the National Slate Museum in Llanberis, North Wales.

The largest working water wheel in the world is the Laxey Wheel (also known as *Lady Isabella*) in the village of Laxey, Isle of Man. It is 72 feet 6 inches (22.10 m) in diameter and 6 feet (1.83 m) wide and is maintained by Manx National Heritage.

Lady Isabella Wheel, Laxey, Isle of Man, used to drive mine pumps

Development of water turbines during the Industrial Revolution led to decreased popularity of water wheels. The main advantage of turbines is that its ability to harness head is much greater than the diameter of the turbine, whereas a water wheel cannot effectively harness head greater than its diameter. The migration from water wheels to modern turbines took about one hundred years.

A mid-19th-century water wheel at Cromford in England used for grinding locally mined barytes.

China

Chinese water wheels almost certainly have a separate origin, as early ones there were invariably horizontal water wheels. By at least the 1st century AD, the Chinese of the Eastern Han Dynasty were using water wheels to crush grain in mills and to power the piston-bellows in forging iron ore into cast iron.

In the text known as the *Xin Lun* written by Huan Tan about 20 AD (during the usurpation of Wang Mang), it states that the legendary mythological king known as Fu Xi was the one responsible for the pestle and mortar, which evolved into the tilt-hammer and then trip hammer device. Although the author speaks of the mythological Fu Xi, a passage of his writing gives hint that the water wheel was in widespread use by the 1st century AD in China.

Fu Hsi invented the pestle and mortar, which is so useful, and later on it was cleverly improved in such a way that the whole weight of the body could be used for treading on the tilt-hammer (*tui*), thus increasing the efficiency ten times. Afterwards the power of animals—donkeys, mules, oxen, and horses—was applied by means of machinery, and water-power too used for pounding, so that the benefit was increased a hundredfold.

In the year 31 AD, the engineer and Prefect of Nanyang, Du Shi (d. 38), applied a complex use of the water wheel and machinery to power the bellows of the blast furnace to create cast iron. Du Shi is mentioned briefly in the *Book of Later Han* (*Hou Han Shu*) as follows (in Wade-Giles spelling):

In the seventh year of the Chien-Wu reign period (31 AD) Tu Shih was posted to be Prefect of

Nanyang. He was a generous man and his policies were peaceful; he destroyed evil-doers and established the dignity (of his office). Good at planning, he loved the common people and wished to save their labor. He invented a water-power reciprocator (*shui phai*) for the casting of (iron) agricultural implements. Those who smelted and cast already had the push-bellows to blow up their charcoal fires, and now they were instructed to use the rushing of the water (*chi shui*) to operate it ... Thus the people got great benefit for little labor. They found the 'water(-powered) bellows' convenient and adopted it widely.

图 4·85 水转翻车
(引自 (明) 宋应星 〖天工开物〗)

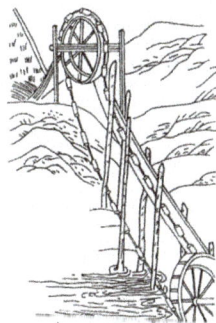

Two types of hydraulic-powered chain pumps from the Tiangong Kaiwu of 1637, written by the Ming Dynasty encyclopedist, Song Yingxing (1587–1666).

Water wheels in China found practical uses such as this, as well as extraordinary use. The Chinese inventor Zhang Heng (78–139) was the first in history to apply motive power in rotating the astronomical instrument of an armillary sphere, by use of a water wheel. The mechanical engineer Ma Jun (c. 200–265) from Cao Wei once used a water wheel to power and operate a large mechanical puppet theater for the Emperor Ming of Wei.

India

The early history of the watermill in India is obscure. Ancient Indian texts dating back to the 4th century BC refer to the term *cakkavattaka* (turning wheel), which commentaries explain as *arahatta-ghati-yanta* (machine with wheel-pots attached). On this basis, Joseph Needham suggested that the machine was a noria. Terry S. Reynolds, however, argues that the "term used in Indian texts is ambiguous and does not clearly indicate a water-powered device." Thorkild Schiøler argued that it is "more likely that these passages refer to some type of tread- or hand-operated water-lifting device, instead of a water-powered water-lifting wheel."

According to Greek historical tradition, India received water-mills from the Roman Empire in the early 4th century AD when a certain Metrodoros introduced "water-mills and baths, unknown among them [the Brahmans] till then". Irrigation water for crops was provided by using water raising wheels, some driven by the force of the current in the river from which the water was being raised. This kind of water raising device was used in ancient India, predating, according to Pacey, its use in the later Roman Empire or China, even though the first literary, archaeological and pictorial evidence of the water wheel appeared in the Hellenistic world.

Around 1150, the astronomer Bhaskara Achārya observed water-raising wheels and imagined such a wheel lifting enough water to replenish the stream driving it, effectively, a perpetual motion machine. The construction of water works and aspects of water technology in India is described in Arabic and Persian works. During medieval times, the diffusion of Indian and Persian irrigation technologies gave rise to an advanced irrigation system which brought about economic growth and also helped in the growth of material culture.

Islamic World

Arab engineers took over the water technology of the hydraulic societies of the ancient Near East; they adopted the Greek water wheel as early as the 7th century, excavation of a canal in the Basra region discovered remains of a water wheel dating from this period. Hama in Syria still preserves some of its large wheels, on the river Orontes, although they are no longer in use. One of the largest had a diameter of about 20 metres and its rim was divided into 120 compartments. Another wheel that is still in operation is found at Murcia in Spain, La Nora, and although the original wheel has been replaced by a steel one, the Moorish system during al-Andalus is otherwise virtually unchanged. Some medieval Islamic compartmented water wheels could lift water as high as 30 meters. Muhammad ibn Zakariya al-Razi's *Kitab al-Hawi* in the 10th century described a noria in Iraq that could lift as much as 153,000 litres per hour, or 2550 litres per minute. This is comparable to the output of modern norias in East Asia, which can lift up to 288,000 litres per hour, or 4800 litres per minute.

The norias of Hama on the Orontes River

The industrial uses of watermills in the Islamic world date back to the 7th century, while horizontal-wheeled and vertical-wheeled water mills were both in widespread use by the 9th century. A variety of industrial watermills were used in the Islamic world, including gristmills, hullers, sawmills, shipmills, stamp mills, steel mills, sugar mills, and tide mills. By the 11th century, every province throughout the Islamic world had these industrial watermills in operation, from al-Andalus and North Africa to the Middle East and Central Asia. Muslim and Christian engineers also used crankshafts and water turbines, gears in watermills and water-raising machines, and dams as a source of water, used to provide additional power to watermills and water-raising machines. Fulling mills and steel mills may have spread from Islamic Spain to Christian Spain in the 12th century. Industrial water mills were also employed in large factory complexes built in al-Andalus between the 11th and 13th centuries.

Water wheel in Djambi, Sumatra, c. 1918

The engineers of the Islamic world developed several solutions to achieve the maximum output from a water wheel. One solution was to mount them to piers of bridges to take advantage of the increased flow. Another solution was the shipmill, a type of water mill powered by water wheels mounted on the sides of ships moored in midstream. This technique was employed along the Tigris and Euphrates rivers in 10th-century Iraq, where large shipmills made of teak and iron could produce 10 tons of flour from corn every day for the granary in Baghdad. The flywheel mechanism, which is used to smooth out the delivery of power from a driving device to a driven machine, was invented by Ibn Bassal (fl. 1038-1075) of Al-Andalus; he pioneered the use of the flywheel in the saqiya (chain pump) and noria. The engineers Al-Jazari in the 13th century and Taqi al-Din in the 16th century described many inventive water-raising machines in their technological treatises. They also employed water wheels to power a variety of devices, including various water clocks and automata.

Types

Most water wheels in the United Kingdom and the United States are (or were) vertical wheels rotating about a horizontal axle, but in the Scottish highlands and parts of southern Europe mills often had a horizontal wheel (with a vertical axle). Water wheels are classified by the way in which water is applied to the wheel, relative to the wheel's axle. Overshot and pitchback water wheels are suitable where there is a small stream with a height difference of more than 2 meters, often in association with a small reservoir. Breastshot and undershot wheels can be used on rivers or high volume flows with large reservoirs.

Horizontal Wheel

Commonly called a tub wheel or Norse mill, the horizontal wheel is essentially a very primitive and inefficient form of the modern turbine. It is usually mounted inside a mill building below the working floor. A jet of water is directed on to the paddles of the water wheel, causing them to turn; water exits beneath the wheel, generally through the center. This is a simple system, usually used without gearing so that the vertical axle of the water wheel becomes the drive spindle of the mill.

Undershot Wheel

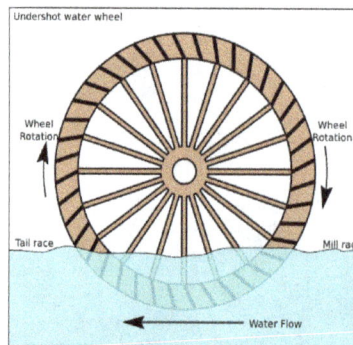

Undershot water wheel

An undershot wheel (also called a *stream wheel*) is a vertically mounted water wheel that is rotated by water striking paddles or blades at the bottom of the wheel. The name *undershot* comes from this striking at the bottom of the wheel. This type of water wheel is the oldest type of wheel.

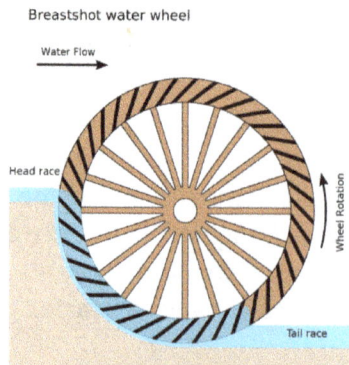

Breastshot water wheel

It is also regarded as the least efficient type, although subtypes of this water wheel (e.g. the Poncelet wheel, Sagebien wheel and Zuppinger wheel) allow somewhat greater efficiencies than the traditional undershot wheels. The advantages of undershot wheels are that they are somewhat cheaper and simpler to build, and have less of an environmental impact—as they do not constitute a major change of the river. Their disadvantages are—as mentioned before—less efficiency, which means that they generate less power and can only be used where the flow rate is sufficient to provide torque.

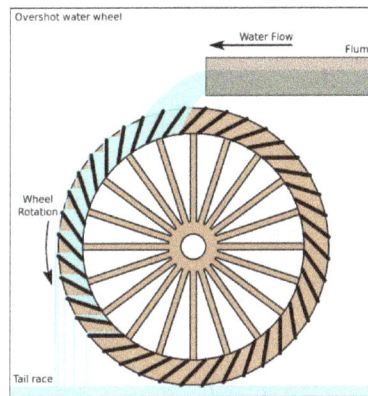

Overshot water wheel

Undershot wheels gain no advantage from head. They are most suited to shallow streams in flat country.

Undershot wheels are also well suited to installation on floating platforms. The earliest were probably constructed by the Byzantine general Belisarius during the siege of Rome in 537. Later they were sometimes mounted immediately downstream from bridges where the flow restriction of arched bridge piers increased the speed of the current.

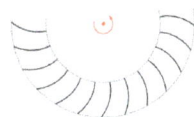

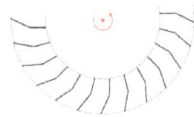

Sabegien, Poncelet and Zuppinger water wheel

Breastshot Wheel

A vertically mounted water wheel that is rotated by falling water striking buckets near the center of the wheel's edge, or just above it, is said to be *breastshot*. Breastshot wheels are the most common type in the United States of America and are said to have powered the American industrial revolution.

Breastshot wheels are less efficient than overshot wheels, are more efficient than undershot wheels, and are not backshot. The individual blades of a breastshot wheel are actually buckets, as are those of most overshot wheels, and not simple paddles like those of most undershot wheels. A breastshot wheel requires a good trash rack and typically has a masonry "apron" closely conforming to the wheel face, which helps contain the water in the buckets as they progress downwards. Breastshot wheels are preferred for steady, high-volume flows such as are found on the fall line of the North American East Coast.

The Anderson Mill of Texas is undershot, backshot, and overshot using two sources of water. This allows the speed of the wheel to be controlled.

Overshot Wheel

A vertically mounted water wheel that is rotated by falling water striking paddles, blades or buckets near the top of the wheel is said to be *overshot*. In true overshot wheels the water passes over the top of the wheel, but the term is sometimes applied to backshot or pitchback wheels where the water goes down behind the water wheel.

A typical overshot wheel has the water channeled to the wheel at the top and slightly beyond the axle. The water collects in the buckets on that side of the wheel, making it heavier than the other "empty" side. The weight turns the wheel, and the water flows out into the tail-water when the wheel rotates enough to invert the buckets. The overshot design can use all of the water flow for power (unless there is a leak) and does not require rapid flow.

Unlike undershot wheels, overshot wheels gain a double advantage from gravity. Not only is the momentum of the flowing water partially transferred to the wheel, the weight of the water descending in the wheel's buckets also imparts additional energy. The mechanical power derived from an overshot wheel is determined by the wheel's physical size and the available head, so they are ideally suited to hilly or mountainous country. On average, the undershot wheel uses 22 percent of the energy in the flow of water, while an overshot wheel uses 63 percent, as calculated by English civil engineer John Smeaton in the 18th century.

Overshot wheels demand exact engineering and significant head, which usually means significant investment in constructing a dam, millpond and waterways. Sometimes the final approach of the water to the wheel is along a lengthy flume or penstock.

Reversible Wheel

A special type of overshot wheel is the reversible water wheel. This has two sets of blades or buckets running in opposite directions, so that it can turn in either direction depending on which side the water is directed. Reversible wheels were used in mining industry in order to power various means of ore conveyance. By changing the direction of the wheel, barrels or baskets of ore could be lifted up or lowered down a shaft. As a rule there was also a cable drum or a chain basket (German: *Kettenkorb*) on the axle of the wheel. It was also essential that the wheel had braking equipment in order to be able to stop the wheel (known as a braking wheel). The oldest known drawing of a reversible water wheel was by Georgius Agricola and dates to 1556.

Replica of a reversible wheel with a 9.5 m diameter in Clausthal-Zellerfeld

Backshot Wheel

A backshot wheel (also called *pitchback*) is a variety of overshot wheel where the water is introduced just behind the summit of the wheel. It combines the advantages from breastshot and overshot systems, since the full amount of the potential energy released by the falling water is harnessed as the water descends the back of the wheel (as in overshot wheel) while it also gains power from the water's current past the bottom of the wheel (as in breastshot wheel).

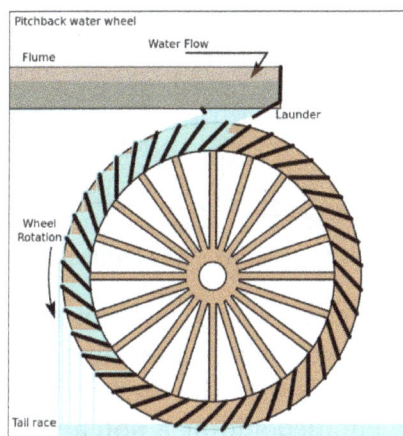

Pitchback or "backshot" water wheel

A backshot wheel continues to function until the water in the wheel pit rises well above the height of the axle, when any other overshot wheel will be stopped or even destroyed. This makes the technique particularly suitable for streams that experience extreme seasonal variations in flow, and reduces the need for complex sluice and tail race configurations.

Backshot wheel at New Lanark World Heritage Site, Scotland

The direction of rotation of a backshot wheel is the same as that of a breastshot wheel at the same location so it can easily replace one, without causing the directional gearing in the mill to be changed. This would increase the power available while only requiring a change to be made to the water level in the top pond, which in some cases is economically viable.

Suspension Wheels and Rim-gears

The suspension wheel with rim-gearing at the Portland Basin Canal Warehouse

Two early improvements were suspension wheels and rim gearing. Suspension wheels are constructed in the same manner as a bicycle wheel, the rim being supported under tension from the hub- this led to larger lighter wheels than the former design where the heavy spokes were under compression. Rim-gearing entailed adding a notched wheel to the rim or shroud of the wheel. A stub gear engaged the rim-gear and took the power into the mill using an independent line shaft. This removed the rotative stress from the axle which could thus be lighter, and also allowed more flexibility in the location of the power train. The shaft rotation was geared up from that of the wheel which led to less power loss. An example of this design pioneered by Thomas Hewes and refined by William Fairburn can be seen at the 1849 restored wheel at the Portland Basin Canal Warehouse.

Efficiency

Overshot (and particularly backshot) wheels are the most efficient type; a backshot steel wheel can be more efficient (about 60%) than all but the most advanced and well-constructed turbines. In some situations an overshot wheel is preferable to a turbine.

The development of the hydraulic turbine wheels with their improved efficiency (>67%) opened up an alternative path for the installation of water wheels in existing mills, or redevelopment of abandoned mills.

Power Calculations

In an undershot wheel or a run of the river wheel the power is dependant to the kinetic energy of the river. Approximate power can be calculated.

Power in Watts= $100 \times A \times V^3 \times C$

A = Area of paddles in the water (square meters)

V = Velocity of the stream in meters per second

C = Efficiency Constant (assume 1 for a water to wire efficiency of 20%)

Rotational speed of the wheel = $9 \times V / D$ rpm

D = diameter in metres

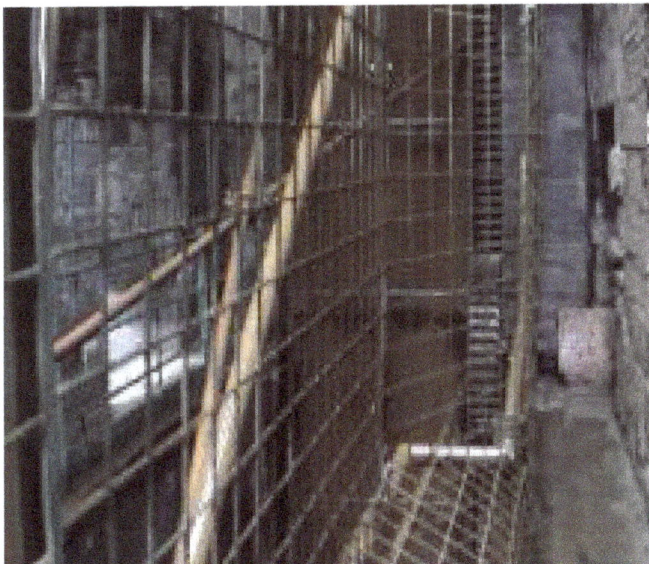

The great water wheel in the Welsh National Slate Museum

For a breast shot or over shot wheel both potential energy and kinetic energy must be considered. This takes the form of the weight of water in the buckets and the vertical distance travelled. A rule of thumb formula is

Power in Watts = 4 × Q × H × C

Q = Weight of water (volume per sec x capacity of the buckets)

V = Velocity of the stream in meters per second

H = Head, or height difference of water between the lip of the flume (head race) and the tailrace

C = Efficiency Constant

The optimal rotational speed of a breast shot or overshot wheel is approximately:

Rotational speed of the wheel= 21/ √D

D = diameter of the wheel in metres

Hydraulic Wheel

A recent development of the breastshot wheel is a hydraulic wheel which effectively incorporates automatic regulation systems. The Aqualienne is one example. It generates between 37 kW and 200 kW of electricity from a $20m^3$ waterflow with a head of 1 to 3.5m. It is designed to produce electricity at the sites of former watermills.

Hydraulic Wheel Part Reaction Turbine

A parallel development is the hydraulic wheel/part reaction turbine that also incorporates a weir into the centre of the wheel but uses blades angled to the water flow. The WICON-Stem Pressure Machine (SPM) exploits this flow. Estimated efficiency 67%.

The University of Southampton School of Civil Engineering and the Environment in the UK has investigated both types of Hydraulic wheel machines and has estimated their hydraulic efficiency and suggested improvements, i.e. The Rotary Hydraulic Pressure Machine. (Estimated maximum efficiency 85%).

These type of water wheels have high efficiency at part loads / variable flows and can operate at very low heads, < 1 metre. Combined with direct drive Axial Flux Permanent Magnet Alternators and power electronics they offer a viable alternative for low head hydroelectric power generation.

Water-lifting

In water-raising devices rotary motion is typically more efficient than machines based on oscillating motion.

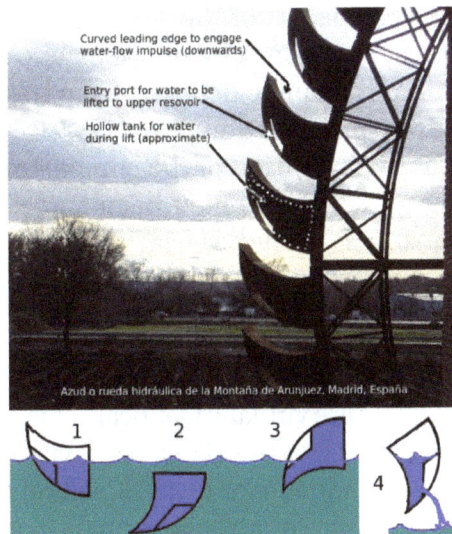

Curved leading edge to engage water-flow impulse (downwards)

Entry port for water to be lifted to upper resovoir

Hollow tank for water during lift (approximate)

Azud o rueda hidráulica de la Montaña de Arunjuez, Madrid, España

Detail of azud at Aranjuez, Spain

The compartmented water wheel comes in two basic forms, the wheel with compartmented body (Latin *tympanum*) and the wheel with compartmented rim or a rim with separate, attached containers. The wheels could be either turned by the flow of water, men treading on its outside or by animals by means of a sakia gear. While the tympanum had a large discharge capacity, it could lift the water only to less than the height of its own radius and required a large torque for rotating. These constructional deficiencies were overcome by the wheel with a compartmented rim which was a less heavy design with a higher lift.

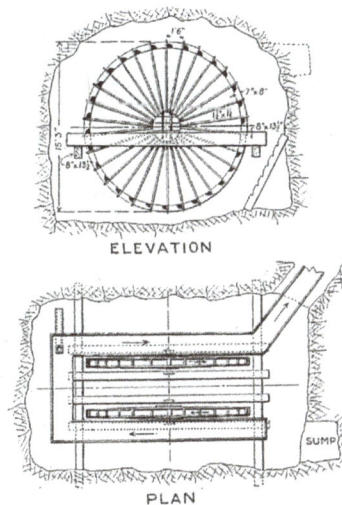

ELEVATION

PLAN

Drainage wheel from Rio Tinto mines

Ptolemaic Egypt

The earliest literary reference to a water-driven, compartmented wheel appears in the technical treatise *Pneumatica* (chap. 61) of the Greek engineer Philo of Byzantium (ca. 280–220 BC). In his *Parasceuastica* (91.43–44), Philo advises the use of such wheels for submerging siege mines as a defensive measure against enemy sapping. Compartmented wheels appear to have been the means of choice for draining dry docks in Alexandria under the reign of Ptolemy IV (221–205 BC). Several Greek papyri of the 3rd to 2nd century BC mention the use of these wheels, but don't give further details. The non-existence of the device in the Ancient Near East before Alexander's conquest can be deduced from its pronounced absence from the otherwise rich oriental iconography on irrigation practices. Unlike other water-lifting devices and pumps of the period though, the invention of the compartmented wheel cannot be traced to any particular Hellenistic engineer and may have been made in the late 4th century BC in a rural context away from the metropolis of Alexandria.

The earliest depiction of a compartmented wheel is from a tomb painting in Ptolemaic Egypt which dates to the 2nd century BC. It shows a pair of yoked oxen driving the wheel via a sakia gear, which is here for the first time attested, too. The Greek sakia gear system is already shown fully developed to the point that "modern Egyptian devices are virtually identical". It is assumed that the scientists of the Museum of Alexandria, at the time the most active Greek research center, may have been involved in its invention. An episode from the Alexandrian War in 48 BC tells of how Caesar's enemies employed geared water wheels to pour sea water from elevated places on the position of the trapped Romans.

Around 300 AD, the noria was finally introduced when the wooden compartments were replaced with inexpensive ceramic pots that were tied to the outside of an open-framed wheel.

Environmental Impact of Reservoirs

The environmental impact of reservoirs comes under ever increasing scrutiny as the global demand for water and energy increases and the number and size of reservoirs increases.

The Wachusett Dam in Clinton, Massachusetts.

Dams and reservoirs can be used to supply drinking water, generate hydroelectric power, increase the water supply for irrigation, provide recreational opportunities, and flood control. However, adverse environmental and sociological impacts have been identified during and after many reservoir constructions. Whether reservoir projects are ultimately beneficial or detrimental to either the environment or surrounding human populations has been debated since the 1960s and likely before then, as well. In 1960 the construction of Llyn Celyn and the flooding of Capel Celyn provoked political uproar which continues to this day. More recently, the construction of Three Gorges Dam and other similar projects throughout Asia, Africa and Latin America have generated considerable environmental and political debate.

Upstream Impacts

Lake Nasser behind the Aswan dam, Egypt, 5250 km², displaced 60,000 people

Fragmentation of river Ecosystems

A dam also acts as a barrier between the upstream and downstream movement of migratory river animals, such as salmon and trout.

Some communities have also begun the practice of transporting migratory fish upstream to spawn via a barge.

Reservoir Sedimentation

Rivers carry sediment down their riverbeds, allowing for the formation of depositional features such as river deltas, alluvial fans, braided rivers, oxbow lakes, levees and coastal shores. The construction of a dam blocks the flow of sediment downstream, leading to downstream erosion of these Sedimentary depositional environments, and increased sediment build-up in the reservoir. While the rate of sedimentation varies for each dam and each river, eventually all reservoirs develop a reduced water-storage capacity due to the exchange of storage space for sediment. Diminished storage capacity results in decreased ability to produce hydroelectric power, reduced availability of water for irrigation, and if left unaddressed, may ultimately result in the expiration of the dam and river.

Impact Below Dam

Riverline and Coastal Erosion

As all dams result in reduced sediment load downstream, a dammed river is said to be "hungry" for sediment. Because the rate of deposition of sediment is greatly reduced since there is less to deposit but the rate of erosion remains nearly constant, the water flow erodes the river shores and riverbed, threatening shoreline ecosystems, deepening the riverbed, and narrowing the river over time. This leads to a compromised water table, reduced water levels, homogenization of the river flow and thus reduced ecosystem variability, reduced support for wildlife, and reduced amount of sediment reaching coastal plains and deltas. This prompts coastal erosion, as beaches are unable to replenish what waves erode without the sediment deposition of supporting river systems. Downsteam channel erosion of dammed rivers is related to the morphology of the riverbed, which is different from directly studying the amounts of sedimentation because it is subject to specific long term conditions for each river system. For example, the eroded channel could create a lower water table level in the affected area, impacting bottomland crops such as alfalfa or corn, and resulting in a smaller supply. In the case of the Three Gorges Dam in China the changes described above now appears to have arrived at a new balance of erosion and sedimentation over a 10-year period in the lower reaches of the river. The impacts on the tidal region have also been linked to the upstream effects of the dam.

Water Temperature

The water of a deep reservoir in temperate climates typically stratifies with a large volume of cold, oxygen poor water in the hypolimnion. Analysis of temperature profiles from 11 large dams in the Murray Darling Basin (Australia) indicated differences between surface water and bottom water temperatures up to 16.7 degrees Celsius. If this water is released to maintain river flow, it can cause adverse impacts on the downstream ecosystem including fish populations. Under worse case conditions (such as when the reservoir is full or near full), the stored water is strongly stratified and large volumes of water are being released to the downstream river channel via bottom level outlets, depressed temperatures can be detected 250 - 350 kilometres downstream. The operators of Burrendong Dam on the Macquarie River (eastern Australia) are attempting to address thermal suppression by hanging a geotextile curtain around the existing outlet tower to force the selective release of surface water.

Effects Beyond the Reservoir

Effects on Humans

Diseases

Whilst reservoirs are helpful to humans, they can also be harmful as well. One negative effect is that the reservoirs can become breeding grounds for disease vectors. This holds true especially in tropical areas where mosquitoes (which are vectors for malaria) and snails (which are vectors for Schistosomiasis) can take advantage of this slow flowing water.

Resettlement

Dams and the creation of reservoirs also require relocation of potentially large human populations if they are constructed close to residential areas. The record for the largest population relocated belongs to the Three Gorges dam built in China. Its reservoir submerged a large area of land, forcing over a million people to relocate. "Dam related relocation affects society in three ways: an economic disaster, human trauma, and social catastrophe", states Dr. Michael Cernea of the World Bank and Dr. Thayer Scudder, a professor at the California Institute of Technology. As well, as resettlement of communities, care must also be taken not to irreparably damage sites of historical or cultural value. The Aswan Dam forced the movement of the Temple at Aswan to prevent its destruction by the flooding of the reservoir.

Lake Manantali, 477 km², displaced 12,000 people.

Disaster

Dams occasionally break causing catastrophic damage to communities downstream. Dams break due to engineering errors, attack or natural disaster. The greatest dam break disaster happened in China killing 200,000 Chinese citizens. However, they have happened in California killing 600 people, Germany during World War II and other countries.

Flood Control

The controversial Three Gorges Dam in China is able to store 22 cubic kilometres of floodwaters on the Yangtze River. The 1954 Yangtze River floods killed 33,000 people and displaced 18 million people from their homes. In 1998 a flood killed 4000 people and 180 million people were affected. A flood in August 2009 was completely captured behind the new dam.

Effects on Flood-dependent Ecology and Agriculture

In many developing countries the savanna and forest ecology of the floodplains depend on seasonal flooding from rivers. Also, flood recession cropping is practiced extensively whereby the land is cultivated taking advantage of the residual soil moisture after floods recede. Dams attenuate floods which may affect the ecology and agriculture seriously.

Water becomes scarce for nomadic pastoralist in Baluchistan due to new dam developments for irrigation.

Case Studies

- The Manatali reservoir formed by the Manantali dam in Mali intersects the migration routes of nomadic pastoralists and destroyed 43000 ha of savannah, probably leading to overgrazing and soil erosion elsewhere. Further, the reservoir destroyed 120 km² of forest. The depletion of groundwater aquifers, which is caused by the suppression of the seasonal flood cycle, is damaging the forests downstream of the dam.

- After the closure of the Kainji Dam in Nigeria, 50 to 70 percent of the downstream area of flood-recession cropping was lost.

Greenhouse Gases

Reservoirs may contribute to changes in the Earth's climate. Warm climate reservoirs generate methane, a greenhouse gas when the reservoirs are stratified, in which the bottom layers are anoxic (i.e. they lack oxygen), leading to degradation of biomass through anaerobic processes. At a dam in Brazil, where the flooded basin is wide and the biomass volume is high the methane produced results in a pollution potential 3.5 times more than an oil-fired power plant would be. A theoretical study has indicated that globally hydroelectric reservoirs may emit 104 million metric tonnes of methane gas annually. Methane gas is a significant contributor to global climate change.

The following table indicates reservoir emissions in milligrams per square meter per day for different bodies of water.

Location	Carbon Dioxide	Methane
Lakes	700	9
Temperate reservoirs	1500	20
Tropical reservoirs	3000	100

References

- Bent Sørensen (2004). Renewable Energy: Its Physics, Engineering, Use, Environmental Impacts, Economy, and Planning Aspects. Academic Press. pp. 556–. ISBN 978-0-12-656153-1.

- Adriana de Miranda (2007). Water architecture in the lands of Syria: the water-wheels. L'Erma di Bretschneider. pp. 38–9. ISBN 88-8265-433-8.

- *Nevell, Mike; Walker (2001). Portland Basin and the archaeology of the Canal Warehouse. Tameside Metropolitan Borough with University of Manchester Archaeological Unit. ISBN 1-871324-25-4.

- Hurford, Clive.; Schneider, M; Cowx, Ian; West, Richard (25 June 1997), "21", Conservation Monitoring in Freshwater Habitats, Berlin: Springer Dordrecht, pp. 219–230, doi:10.1007/978-1-4020-9278-7, ISBN 978-1-4020-9277-0, ISSN 0343-6993

- William R. Jobin, 1999. Dams and Disease: Ecological Design and Health Impacts of Large Dams, Canals, and Irrigation Systems, Taylor & Francis, ISBN 0-419-22360-6

- Thomson, Ross (2009). Structures of Change in the Mechanical Age: Technological Invention in the United States 1790-1865. Baltimore, MD: The Johns Hopkins University Press. p. 34. ISBN 978-0-8018-9141-0.

- Hunt, Robert (1887). British Mining: A Treatise in the History, Discovery, Practical Development, and Future Prospects of Metalliferous Mines of the United Kingdom (2nd ed.). London: Crosby Lockwood and Co. p. 505. Retrieved 2 May 2015.

- Al-Hassani, Salim. "800 Years Later: In Memory of Al-Jazari, A Genius Mechanical Engineer". Muslim Heritage. The Foundation for Science, Technology, and Civilisation. Retrieved 30 April 2015.

- Rabl A. et. al. (August 2005). "Final Technical Report, Version 2" (PDF). Externalities of Energy: Extension of Accounting Framework and Policy Applications. European Commission. Archived from the original (PDF) on March 7, 2012.

- Kreis, Steven (2001). "The Origins of the Industrial Revolution in England". The history guide. Retrieved 19 June 2010.

4

Fundamentals of Solar Energy

Solar energy is light and heat from the sun that is utilized using a range of technology, such as solar energy, solar thermal energy, solar power and concentrated solar power. Solar energy is an important source of renewable energy and can be broadly characterized as either passive solar or active solar.

Solar Energy

Solar energy is radiant light and heat from the Sun that is harnessed using a range of ever-evolving technologies such as solar heating, photovoltaics, solar thermal energy, solar architecture and artificial photosynthesis.

It is an important source of renewable energy and its technologies are broadly characterized as either passive solar or active solar depending on how they capture and distribute solar energy or convert it into solar power. Active solar techniques include the use of photovoltaic systems, concentrated solar power and solar water heating to harness the energy. Passive solar techniques include orienting a building to the Sun, selecting materials with favorable thermal mass or light-dispersing properties, and designing spaces that naturally circulate air.

The large magnitude of solar energy available makes it a highly appealing source of electricity. The United Nations Development Programme in its 2000 World Energy Assessment found that the annual potential of solar energy was 1,575–49,837 exajoules (EJ). This is several times larger than the total world energy consumption, which was 559.8 EJ in 2012.

In 2011, the International Energy Agency said that "the development of affordable, inexhaustible and clean solar energy technologies will have huge longer-term benefits. It will increase countries' energy security through reliance on an indigenous, inexhaustible and mostly import-independent resource, enhance sustainability, reduce pollution, lower the costs of mitigating global warming, and keep fossil fuel prices lower than otherwise. These advantages are global. Hence the additional costs of the incentives for early deployment should be considered learning investments; they must be wisely spent and need to be widely shared".

Potential

The Earth receives 174,000 terawatts (TW) of incoming solar radiation (insolation) at the upper atmosphere. Approximately 30% is reflected back to space while the rest is absorbed by clouds, oceans and land masses. The spectrum of solar light at the Earth's surface is mostly spread across the visible and near-infrared ranges with a small part in the near-ultraviolet. Most of the world's population live in areas with insolation levels of 150-300 watts/m², or 3.5-7.0 kWh/m² per day.

Solar radiation is absorbed by the Earth's land surface, oceans – which cover about 71% of the globe – and atmosphere. Warm air containing evaporated water from the oceans rises, causing atmospheric circulation or convection. When the air reaches a high altitude, where the temperature is low, water vapor condenses into clouds, which rain onto the Earth's surface, completing the water cycle. The latent heat of water condensation amplifies convection, producing atmospheric phenomena such as wind, cyclones and anti-cyclones. Sunlight absorbed by the oceans and land masses keeps the surface at an average temperature of 14 °C. By photosynthesis, green plants convert solar energy into chemically stored energy, which produces food, wood and the biomass from which fossil fuels are derived.

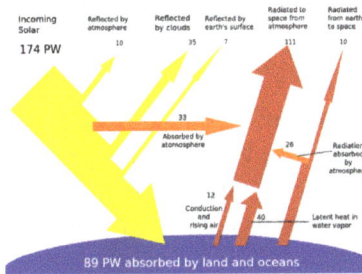

About half the incoming solar energy reaches the Earth's surface.

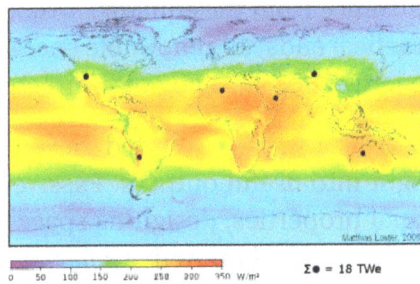

Average insolation. The theoretical area of the small black dots is sufficient to supply the world's total energy needs of 18 TW with solar power.

The total solar energy absorbed by Earth's atmosphere, oceans and land masses is approximately 3,850,000 exajoules (EJ) per year. In 2002, this was more energy in one hour than the world used in one year. Photosynthesis captures approximately 3,000 EJ per year in biomass. The amount of solar energy reaching the surface of the planet is so vast that in one year it is about twice as much as will ever be obtained from all of the Earth's non-renewable resources of coal, oil, natural gas, and mined uranium combined,

Yearly solar luxes & human consumption	
Solar	3,850,000
Wind	2,250
Biomass potential	~200
Primary energy use2	539
Electricity2	~67
1 Energy given in Exajoule (EJ) = 1018 J = 278 TWh 2 Consumption as of year 2010	

The potential solar energy that could be used by humans differs from the amount of solar energy present near the surface of the planet because factors such as geography, time variation, cloud cover, and the land available to humans limit the amount of solar energy that we can acquire.

Geography affects solar energy potential because areas that are closer to the equator have a greater amount of solar radiation. However, the use of photovoltaics that can follow the position of the sun can significantly increase the solar energy potential in areas that are farther from the equator. Time variation effects the potential of solar energy because during the nighttime there is little solar radiation on the surface of the Earth for solar panels to absorb. This limits the amount of energy that solar panels can absorb in one day. Cloud cover can affect the potential of solar panels because clouds block incoming light from the sun and reduce the light available for solar cells.

In addition, land availability has a large effect on the available solar energy because solar panels can only be set up on land that is otherwise unused and suitable for solar panels. Roofs have been found to be a suitable place for solar cells, as many people have discovered that they can collect energy directly from their homes this way. Other areas that are suitable for solar cells are lands that are not being used for businesses where solar plants can be established.

Solar technologies are characterized as either passive or active depending on the way they capture, convert and distribute sunlight and enable solar energy to be harnessed at different levels around the world, mostly depending on distance from the equator. Although solar energy refers primarily to the use of solar radiation for practical ends, all renewable energies, other than Geothermal power and Tidal power, derive their energy either directly or indirectly from the Sun.

Active solar techniques use photovoltaics, concentrated solar power, solar thermal collectors, pumps, and fans to convert sunlight into useful outputs. Passive solar techniques include selecting materials with favorable thermal properties, designing spaces that naturally circulate air, and referencing the position of a building to the Sun. Active solar technologies increase the supply of energy and are considered supply side technologies, while passive solar technologies reduce the need for alternate resources and are generally considered demand side technologies.

In 2000, the United Nations Development Programme, UN Department of Economic and Social Affairs, and World Energy Council published an estimate of the potential solar energy that could be used by humans each year that took into account factors such as insolation, cloud cover, and the land that is usable by humans. The estimate found that solar energy has a global potential of 1,575–49,837 EJ per year *(see table below)*.

Annual solar energy potential by region (Exajoules)											
Region	North America	Latin America and Caribbean	Western Europe	Central and Eastern Europe	Former Soviet Union	Middle East and North Africa	Sub-Saharan Africa	Pacific Asia	South Asia	Centrally planned Asia	Pacific OECD
Minimum	181.1	112.6	25.1	4.5	199.3	412.4	371.9	41.0	38.8	115.5	72.6
Maximum	7,410	3,385	914	154	8,655	11,060	9,528	994	1,339	4,135	2,263

Thermal Energy

Solar thermal technologies can be used for water heating, space heating, space cooling and process heat generation.

Early Commercial Adaptation

In 1897, Frank Shuman, a U.S. inventor, engineer and solar energy pioneer built a small demonstration solar engine that worked by reflecting solar energy onto square boxes filled with ether, which has a lower boiling point than water, and were fitted internally with black pipes which in turn powered a steam engine. In 1908 Shuman formed the Sun Power Company with the intent of building larger solar power plants. He, along with his technical advisor A.S.E. Ackermann and British physicist Sir Charles Vernon Boys, developed an improved system using mirrors to reflect solar energy upon collector boxes, increasing heating capacity to the extent that water could now be used instead of ether. Shuman then constructed a full-scale steam engine powered by low-pressure water, enabling him to patent the entire solar engine system by 1912.

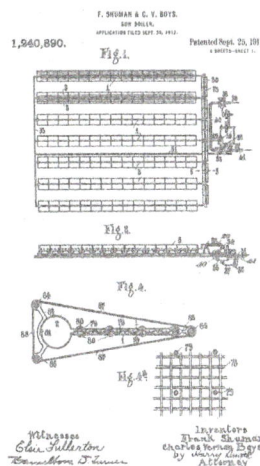

1917 Patent drawing of Shuman's solar collector

Shuman built the world's first solar thermal power station in Maadi, Egypt, between 1912 and 1913. His plant used parabolic troughs to power a 45–52 kilowatts (60–70 hp) engine that pumped more than 22,000 litres (4,800 imp gal; 5,800 US gal) of water per minute from the Nile River to adjacent cotton fields. Although the outbreak of World War I and the discovery of cheap oil in the 1930s discouraged the advancement of solar energy, Shuman's vision and basic design were resurrected in the 1970s with a new wave of interest in solar thermal energy. In 1916 Shuman was quoted in the media advocating solar energy's utilization, saying:

We have proved the commercial profit of sun power in the tropics and have more particularly proved that after our stores of oil and coal are exhausted the human race can receive unlimited power from the rays of the sun.

Water Heating

Solar hot water systems use sunlight to heat water. In low geographical latitudes (below 40 degrees) from 60 to 70% of the domestic hot water use with temperatures up to 60 °C can be provided by solar heating systems. The most common types of solar water heaters are evacuated tube collectors (44%) and glazed flat plate collectors (34%) generally used for domestic hot water; and unglazed plastic collectors (21%) used mainly to heat swimming pools.

Solar water heaters facing the Sun to maximize gain

As of 2007, the total installed capacity of solar hot water systems was approximately 154 thermal gigawatt (GW$_{th}$). China is the world leader in their deployment with 70 GW$_{th}$ installed as of 2006 and a long-term goal of 210 GW$_{th}$ by 2020. Israel and Cyprus are the per capita leaders in the use of solar hot water systems with over 90% of homes using them. In the United States, Canada, and Australia, heating swimming pools is the dominant application of solar hot water with an installed capacity of 18 GW$_{th}$ as of 2005.

Heating, Cooling and Ventilation

In the United States, heating, ventilation and air conditioning (HVAC) systems account for 30% (4.65 EJ/yr) of the energy used in commercial buildings and nearly 50% (10.1 EJ/yr) of the energy used in residential buildings. Solar heating, cooling and ventilation technologies can be used to offset a portion of this energy.

Thermal mass is any material that can be used to store heat—heat from the Sun in the case of solar energy. Common thermal mass materials include stone, cement and water. Historically they have been used in arid climates or warm temperate regions to keep buildings cool by absorbing solar energy during the day and radiating stored heat to the cooler atmosphere at night. However, they can be used in cold temperate areas to maintain warmth as well. The size and placement of thermal mass depend on several factors such as climate, daylighting and shading conditions. When properly incorporated, thermal mass maintains space temperatures in a comfortable range and reduces the need for auxiliary heating and cooling equipment.

MIT's Solar House #1, built in 1939 in the U.S., used seasonal thermal energy storage for year-round heating.

A solar chimney (or thermal chimney, in this context) is a passive solar ventilation system composed of a vertical shaft connecting the interior and exterior of a building. As the chimney warms, the air inside is heated causing an updraft that pulls air through the building. Performance can be improved by using glazing and thermal mass materials in a way that mimics greenhouses.

Deciduous trees and plants have been promoted as a means of controlling solar heating and cooling. When planted on the southern side of a building in the northern hemisphere or the northern side in the southern hemisphere, their leaves provide shade during the summer, while the bare limbs allow light to pass during the winter. Since bare, leafless trees shade 1/3 to 1/2 of incident solar radiation, there is a balance between the benefits of summer shading and the corresponding loss of winter heating. In climates with significant heating loads, deciduous trees should not be planted on the Equator-facing side of a building because they will interfere with winter solar availability. They can, however, be used on the east and west sides to provide a degree of summer shading without appreciably affecting winter solar gain.

Cooking

Solar cookers use sunlight for cooking, drying and pasteurization. They can be grouped into three broad categories: box cookers, panel cookers and reflector cookers. The simplest solar cooker is the box cooker first built by Horace de Saussure in 1767. A basic box cooker consists of an insulated container with a transparent lid. It can be used effectively with partially overcast skies and will typically reach temperatures of 90–150 °C (194–302 °F). Panel cookers use a reflective panel to direct sunlight onto an insulated container and reach temperatures comparable to box cookers. Reflector cookers use various concentrating geometries (dish, trough, Fresnel mirrors) to focus light on a cooking container. These cookers reach temperatures of 315 °C (599 °F) and above but require direct light to function properly and must be repositioned to track the Sun.

Parabolic dish produces steam for cooking, in Auroville, India

Process Heat

Solar concentrating technologies such as parabolic dish, trough and Scheffler reflectors can provide process heat for commercial and industrial applications. The first commercial system was the Solar Total Energy Project (STEP) in Shenandoah, Georgia, USA where a field of 114 parabolic dishes provided 50% of the process heating, air conditioning and electrical requirements for a clothing factory. This grid-connected cogeneration system provided 400 kW of electricity plus thermal energy in the form of 401 kW steam and 468 kW chilled water, and had a one-hour peak load thermal storage. Evaporation ponds are shallow pools that concentrate dissolved solids through evaporation. The use of evaporation ponds to obtain salt from seawater is one of the oldest applications of solar energy. Modern uses include concentrating brine solutions used in leach mining and removing dissolved solids from waste streams. Clothes lines, clotheshorses, and clothes racks dry clothes through evaporation by wind and sunlight without consuming electricity or gas. In some states of the United States legislation protects the "right to dry" clothes. Unglazed transpired collectors (UTC) are perforated sun-facing walls used for preheating ventilation air. UTCs can raise the incoming air temperature up to 22 °C (40 °F) and deliver outlet temperatures of 45–60 °C (113–140 °F). The short payback period of transpired collectors (3 to 12 years) makes them a more cost-effective alternative than glazed collection systems. As of 2003, over 80 systems with a combined collector area of 35,000 square metres (380,000 sq ft) had been installed worldwide, including an 860 m² (9,300 sq ft) collector in Costa Rica used for drying coffee beans and a 1,300 m² (14,000 sq ft) collector in Coimbatore, India, used for drying marigolds.

Water Treatment

Solar distillation can be used to make saline or brackish water potable. The first recorded instance of this was by 16th-century Arab alchemists. A large-scale solar distillation project was first constructed in 1872 in the Chilean mining town of Las Salinas. The plant, which had solar collection area of 4,700 m² (51,000 sq ft), could produce up to 22,700 L (5,000 imp gal; 6,000 US gal) per day and operate for 40 years. Individual still designs include single-slope, double-slope (or greenhouse type), vertical, conical, inverted absorber, multi-wick, and multiple effect. These stills can operate in passive, active, or hybrid modes. Double-slope stills are the most economical for decentralized domestic purposes, while active multiple effect units are more suitable for large-scale applications.

Solar water disinfection in Indonesia

Solar water disinfection (SODIS) involves exposing water-filled plastic polyethylene terephthalate (PET) bottles to sunlight for several hours. Exposure times vary depending on weather and climate from a minimum of six hours to two days during fully overcast conditions. It is recommended by the World Health Organization as a viable method for household water treatment and safe storage. Over two million people in developing countries use this method for their daily drinking water.

Solar energy may be used in a water stabilization pond to treat waste water without chemicals or electricity. A further environmental advantage is that algae grow in such ponds and consume carbon dioxide in photosynthesis, although algae may produce toxic chemicals that make the water unusable.

Electricity Production

Solar power is the conversion of sunlight into electricity, either directly using photovoltaics (PV), or indirectly using concentrated solar power (CSP). CSP systems use lenses or mirrors and tracking systems to focus a large area of sunlight into a small beam. PV converts light into electric current using the photoelectric effect.

Solar power is anticipated to become the world's largest source of electricity by 2050, with solar photovoltaics and concentrated solar power contributing 16 and 11 percent to the global overall consumption, respectively.

Some of the world's largest solar power stations: Ivanpah (CSP) and Topaz (PV)

Commercial CSP plants were first developed in the 1980s. Since 1985 the eventually 354 MW SEGS CSP installation, in the Mojave Desert of California, is the largest solar power plant in the world. Other large CSP plants include the 150 MW Solnova Solar Power Station and the 100 MW Andasol solar power station, both in Spain. The 250 MW Agua Caliente Solar Project, in the United States, and the 221 MW Charanka Solar Park in India, are the world's largest photovoltaic plants. Solar projects exceeding 1 GW are being developed, but most of the deployed photovoltaics are in small rooftop arrays of less than 5 kW, which are connected to the grid using net metering and/or a feed-in tariff. In 2013 solar generated less than 1% of the world's total grid electricity.

Photovoltaics

In the last two decades, photovoltaics (PV), also known as solar PV, has evolved from a pure niche market of small scale applications towards becoming a mainstream electricity source. A solar cell is a device that converts light directly into electricity using the photoelectric effect. The first solar cell was constructed by Charles Fritts in the 1880s. In 1931 a German engineer, Dr Bruno Lange, developed a photo cell using silver selenide in place of copper oxide. Although the prototype selenium cells converted less than 1% of incident light into electricity, both Ernst Werner von Siemens and James Clerk Maxwell recognized the importance of this discovery. Following the work of Russell Ohl in the 1940s, researchers Gerald Pearson, Calvin Fuller and Daryl Chapin created the crystalline silicon solar cell in 1954. These early solar cells cost 286 USD/watt and reached efficiencies of 4.5–6%. By 2012 available efficiencies exceeded 20%, and the maximum efficiency of research photovoltaics was in excess of 40%.

Concentrated Solar Power

Concentrating Solar Power (CSP) systems use lenses or mirrors and tracking systems to focus a large area of sunlight into a small beam. The concentrated heat is then used as a heat source for a conventional power plant. A wide range of concentrating technologies exists; the most developed are the parabolic trough, the concentrating linear fresnel reflector, the Stirling dish and the solar power tower. Various techniques are used to track the Sun and focus light. In all of these systems a working fluid is heated by the concentrated sunlight, and is then used for power generation or energy storage.

Architecture and Urban Planning

Sunlight has influenced building design since the beginning of architectural history. Advanced solar architecture and urban planning methods were first employed by the Greeks and Chinese, who oriented their buildings toward the south to provide light and warmth.

The common features of passive solar architecture are orientation relative to the Sun, compact proportion (a low surface area to volume ratio), selective shading (overhangs) and thermal mass. When these features are tailored to the local climate and environment they can produce well-lit spaces that stay in a comfortable temperature range. Socrates' Megaron House is a classic example of passive solar design. The most recent approaches to solar design use computer modeling tying together solar lighting, heating and ventilation systems in an integrated solar design package. Active solar equipment such as pumps, fans and switchable windows can complement passive design and improve system performance.

Darmstadt University of Technology, Germany, won the 2007 Solar Decathlon in Washington, D.C. with this passive house designed for humid and hot subtropical climate.

Urban heat islands (UHI) are metropolitan areas with higher temperatures than that of the surrounding environment. The higher temperatures result from increased absorption of solar energy by urban materials such as asphalt and concrete, which have lower albedos and higher heat capacities than those in the natural environment. A straightforward method of counteracting the UHI effect is to paint buildings and roads white, and to plant trees in the area. Using these methods, a hypothetical "cool communities" program in Los Angeles has projected that urban temperatures could be reduced by approximately 3 °C at an estimated cost of US$1 billion, giving estimated total annual benefits of US$530 million from reduced air-conditioning costs and healthcare savings.

Agriculture and Horticulture

Agriculture and horticulture seek to optimize the capture of solar energy in order to optimize the productivity of plants. Techniques such as timed planting cycles, tailored row orientation, staggered heights between rows and the mixing of plant varieties can improve crop yields. While sunlight is generally considered a plentiful resource, the exceptions highlight the importance of solar energy to agriculture. During the short growing seasons of the Little Ice Age, French and English farmers employed fruit walls to maximize the collection of solar energy. These walls acted as thermal masses and accelerated ripening by keeping plants warm. Early fruit walls were built perpendicular to the ground and facing south, but over time, sloping walls were developed to make better use of sunlight. In 1699, Nicolas Fatio de Duillier even suggested using a tracking mechanism which could pivot to follow the Sun. Applications of solar energy in agriculture aside from growing crops include pumping water, drying crops, brooding chicks and drying chicken manure. More recently the technology has been embraced by vintners, who use the energy generated by solar panels to power grape presses.

Greenhouses like these in the Westland municipality of the Netherlands grow vegetables, fruits and flowers.

Greenhouses convert solar light to heat, enabling year-round production and the growth (in enclosed environments) of specialty crops and other plants not naturally suited to the local climate. Primitive greenhouses were first used during Roman times to produce cucumbers year-round for the Roman emperor Tiberius. The first modern greenhouses were built in Europe in the 16th century to keep exotic plants brought back from explorations abroad. Greenhouses remain an important part of horticulture today, and plastic transparent materials have also been used to similar effect in polytunnels and row covers.

Transport

Development of a solar-powered car has been an engineering goal since the 1980s. The World Solar Challenge is a biannual solar-powered car race, where teams from universities and enterprises compete over 3,021 kilometres (1,877 mi) across central Australia from Darwin to Adelaide. In 1987, when it was founded, the winner's average speed was 67 kilometres per hour (42 mph) and by 2007 the winner's average speed had improved to 90.87 kilometres per hour (56.46 mph). The North American Solar Challenge and the planned South African Solar Challenge are comparable competitions that reflect an international interest in the engineering and development of solar powered vehicles.

Some vehicles use solar panels for auxiliary power, such as for air conditioning, to keep the interior cool, thus reducing fuel consumption.

In 1975, the first practical solar boat was constructed in England. By 1995, passenger boats incorporating PV panels began appearing and are now used extensively. In 1996, Kenichi Horie made the first solar-powered crossing of the Pacific Ocean, and the *Sun21* catamaran made the first solar-powered crossing of the Atlantic Ocean in the winter of 2006–2007. There were plans to circumnavigate the globe in 2010.

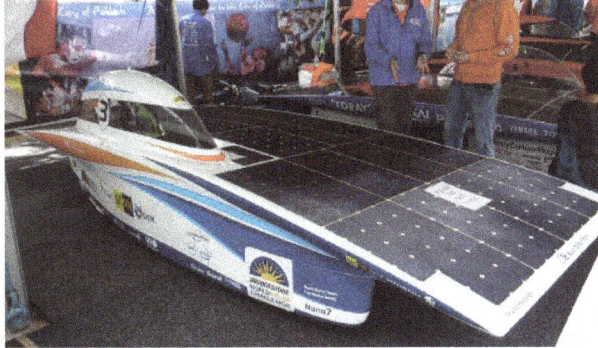

Winner of the 2013 World Solar Challenge in Australia

In 1974, the unmanned AstroFlight Sunrise airplane made the first solar flight. On 29 April 1979, the *Solar Riser* made the first flight in a solar-powered, fully controlled, man-carrying flying machine, reaching an altitude of 40 feet (12 m). In 1980, the *Gossamer Penguin* made the first piloted flights powered solely by photovoltaics. This was quickly followed by the *Solar Challenger* which crossed the English Channel in July 1981. In 1990 Eric Scott Raymond in 21 hops flew from California to North Carolina using solar power. Developments then turned back to unmanned aerial vehicles (UAV) with the *Pathfinder* (1997) and subsequent designs, culminating in the *Helios* which set the altitude record for a non-rocket-propelled aircraft at 29,524 metres (96,864 ft) in 2001. The *Zephyr*, developed by BAE Systems, is the latest in a line of record-breaking solar aircraft, making a 54-hour flight in 2007, and month-long flights were envisioned by 2010. As of 2016, Solar Impulse, an electric aircraft, is currently circumnavigating the globe. It is a single-seat plane powered by solar cells and capable of taking off under its own power. The design allows the aircraft to remain airborne for several days.

Solar electric aircraft circumnavigating the globe in 2015

A solar balloon is a black balloon that is filled with ordinary air. As sunlight shines on the balloon, the air inside is heated and expands causing an upward buoyancy force, much like an artificially heated hot air balloon. Some solar balloons are large enough for human flight, but usage is generally limited to the toy market as the surface-area to payload-weight ratio is relatively high.

Fuel Production

Solar chemical processes use solar energy to drive chemical reactions. These processes offset energy that would otherwise come from a fossil fuel source and can also convert solar energy into storable and transportable fuels. Solar induced chemical reactions can be divided into thermochemical or photochemical. A variety of fuels can be produced by artificial photosynthesis. The multielectron catalytic chemistry involved in making carbon-based fuels (such as methanol) from reduction of carbon dioxide is challenging; a feasible alternative is hydrogen production from protons, though use of water as the source of electrons (as plants do) requires mastering the multielectron oxidation of two water molecules to molecular oxygen. Some have envisaged working solar fuel plants in coastal metropolitan areas by 2050 – the splitting of sea water providing hydrogen to be run through adjacent fuel-cell electric power plants and the pure water by-product going directly into the municipal water system. Another vision involves all human structures covering the earth's surface (i.e., roads, vehicles and buildings) doing photosynthesis more efficiently than plants.

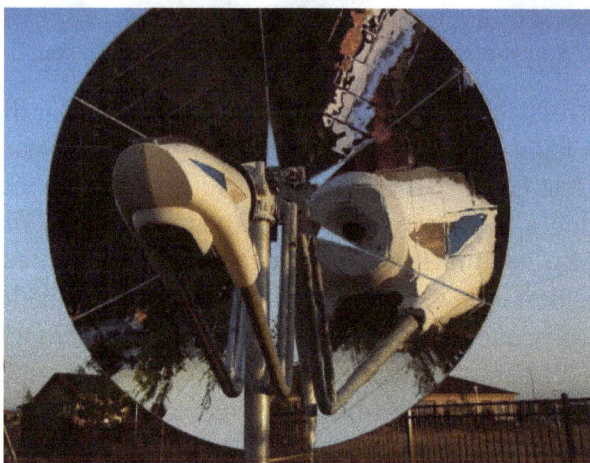

Concentrated solar panels are getting a power boost. Pacific Northwest National Laboratory (PNNL) will be testing a new concentrated solar power system -- one that can help natural gas power plants reduce their fuel usage by up to 20 percent.

Hydrogen production technologies have been a significant area of solar chemical research since the 1970s. Aside from electrolysis driven by photovoltaic or photochemical cells, several thermochemical processes have also been explored. One such route uses concentrators to split water into oxygen and hydrogen at high temperatures (2,300–2,600 °C or 4,200–4,700 °F). Another approach uses the heat from solar concentrators to drive the steam reformation of natural gas thereby increasing the overall hydrogen yield compared to conventional reforming methods. Thermochemical cycles characterized by the decomposition and regeneration of reactants present another avenue for hydrogen production. The Solzinc process under development at the Weizmann Institute of Science uses a 1 MW solar furnace to decompose zinc oxide (ZnO) at temperatures above 1,200 °C (2,200 °F). This initial reaction produces pure zinc, which can subsequently be reacted with water to produce hydrogen.

Energy Storage Methods

Thermal mass systems can store solar energy in the form of heat at domestically useful temperatures for daily or interseasonal durations. Thermal storage systems generally use readily available materials with high specific heat capacities such as water, earth and stone. Well-designed systems can lower peak demand, shift time-of-use to off-peak hours and reduce overall heating and cooling requirements.

Thermal energy storage. The Andasol CSP plant uses tanks of molten salt to store solar energy.

Phase change materials such as paraffin wax and Glauber's salt are another thermal storage medium. These materials are inexpensive, readily available, and can deliver domestically useful temperatures (approximately 64 °C or 147 °F). The "Dover House" (in Dover, Massachusetts) was the first to use a Glauber's salt heating system, in 1948. Solar energy can also be stored at high temperatures using molten salts. Salts are an effective storage medium because they are low-cost, have a high specific heat capacity and can deliver heat at temperatures compatible with conventional power systems. The Solar Two project used this method of energy storage, allowing it to store 1.44 terajoules (400,000 kWh) in its 68 m³ storage tank with an annual storage efficiency of about 99%.

Off-grid PV systems have traditionally used rechargeable batteries to store excess electricity. With grid-tied systems, excess electricity can be sent to the transmission grid, while standard grid electricity can be used to meet shortfalls. Net metering programs give household systems a credit for any electricity they deliver to the grid. This is handled by 'rolling back' the meter whenever the home produces more electricity than it consumes. If the net electricity use is below zero, the utility then rolls over the kilowatt hour credit to the next month. Other approaches involve the use of two meters, to measure electricity consumed vs. electricity produced. This is less common due to the increased installation cost of the second meter. Most standard meters accurately measure in both directions, making a second meter unnecessary.

Pumped-storage hydroelectricity stores energy in the form of water pumped when energy is available from a lower elevation reservoir to a higher elevation one. The energy is recovered when demand is high by releasing the water, with the pump becoming a hydroelectric power generator.

Development, Deployment and Economics

Beginning with the surge in coal use which accompanied the Industrial Revolution, energy consumption has steadily transitioned from wood and biomass to fossil fuels. The early development of solar technologies starting in the 1860s was driven by an expectation that coal would soon become scarce. However, development of solar technologies stagnated in the early 20th century in the face of the increasing availability, economy, and utility of coal and petroleum.

The 1973 oil embargo and 1979 energy crisis caused a reorganization of energy policies around the world and brought renewed attention to developing solar technologies. Deployment strategies focused on incentive programs such as the Federal Photovoltaic Utilization Program in the U.S. and the Sunshine Program in Japan. Other efforts included the formation of research facilities in the U.S. (SERI, now NREL), Japan (NEDO), and Germany (Fraunhofer Institute for Solar Energy Systems ISE).

Participants in a workshop on sustainable development inspect solar panels at Monterrey Institute of Technology and Higher Education, Mexico City on top of a building on campus.

Commercial solar water heaters began appearing in the United States in the 1890s. These systems saw increasing use until the 1920s but were gradually replaced by cheaper and more reliable heating fuels. As with photovoltaics, solar water heating attracted renewed attention as a result of the oil crises in the 1970s but interest subsided in the 1980s due to falling petroleum prices. Development in the solar water heating sector progressed steadily throughout the 1990s and annual growth rates have averaged 20% since 1999. Although generally underestimated, solar water heating and cooling is by far the most widely deployed solar technology with an estimated capacity of 154 GW as of 2007.

The International Energy Agency has said that solar energy can make considerable contributions to solving some of the most urgent problems the world now faces:

The development of affordable, inexhaustible and clean solar energy technologies will have huge longer-term benefits. It will increase countries' energy security through reliance on an indigenous, inexhaustible and mostly import-independent resource, enhance sustainability, reduce pollution, lower the costs of mitigating climate change, and keep fossil fuel prices lower than otherwise. These advantages are global. Hence the additional costs of the incentives for early deployment should be considered learning investments; they must be wisely spent and need to be widely shared.

In 2011, a report by the International Energy Agency found that solar energy technologies such as photovoltaics, solar hot water and concentrated solar power could provide a third of the world's energy by 2060 if politicians commit to limiting climate change. The energy from the sun could play a key role in de-carbonizing the global economy alongside improvements in energy efficiency and imposing costs on greenhouse gas emitters. "The strength of solar is the incredible variety and flexibility of applications, from small scale to big scale".

We have proved ... that after our stores of oil and coal are exhausted the human race can receive unlimited power from the rays of the sun.

ISO Standards

The International Organization for Standardization has established several standards relating to solar energy equipment. For example, ISO 9050 relates to glass in building while ISO 10217 relates to the materials used in solar water heaters.

Solar Thermal Energy

Solar thermal energy (STE) is a form of energy and a technology for harnessing solar energy to generate thermal energy or electrical energy for use in industry, and in the residential and commercial sectors.

Roof-mounted close-coupled thermosiphon solar water heater.

The first three units of Solnova in the foreground, with the two towers of the PS10 and PS20 solar power stations in the background.

Overview

Solar thermal collectors are classified by the United States Energy Information Administration as low-, medium-, or high-temperature collectors. Low-temperature collectors are flat plates generally used to heat swimming pools. Medium-temperature collectors are also usually flat plates but are used for heating water or air for residential and commercial use. High-temperature collectors concentrate sunlight using mirrors or lenses and are generally used for fulfilling heat requirements up to 300 deg C / 20 bar pressure in industries, and for electric power production. Two categories include Concentrated Solar Thermal (CST) for fulfilling heat requirements in industries, and Concentrated Solar Power (CSP) when the heat collected is used for power generation. CST and CSP are not replaceable in terms of application. The largest facilities are located in the American Mojave Desert of California and Nevada. These plants employ a variety of different technologies. The largest examples include, Ivanpah Solar Power Facility (377 MW), Solar Energy Generating Systems installation (354 MW), and Crescent Dunes (110 MW). Spain is the other major developer of solar thermal power plant. The largest examples include, Solnova Solar Power Station (150 MW), the Andasol solar power station (150 MW), and Extresol Solar Power Station (100 MW).

History

Augustin Mouchot demonstrated a solar collector with a cooling engine making ice cream at the 1878 Universal Exhibition in Paris. The first installation of solar thermal energy equipment occurred in the Sahara approximately in 1910 by Frank Shuman when a steam engine was run on steam produced by sunlight. Because liquid fuel engines were developed and found more convenient, the Sahara project was abandoned, only to be revisited several decades later.

Low-temperature Solar Heating and Cooling Systems

Systems for utilizing low-temperature solar thermal energy include means for heat collection; usually heat storage, either short-term or interseasonal; and distribution within a structure or a district heating network. In some cases more than one of these functions is inherent to a single feature of the system (e.g. some kinds of solar collectors also store heat). Some systems are passive, others are active (requiring other external energy to function).

Heating is the most obvious application, but solar cooling can be achieved for a building or district cooling network by using a heat-driven absorption or adsorption chiller (heat pump). There is a productive coincidence that the greater the driving heat from insulation, the greater the cooling output. In 1878, Auguste Mouchout pioneered solar cooling by making ice using a solar steam engine attached to a refrigeration device.

In the United States, heating, ventilation, and air conditioning (HVAC) systems account for over 25% (4.75 EJ) of the energy used in commercial buildings and nearly half (10.1 EJ) of the energy used in residential buildings. Solar heating, cooling, and ventilation technologies can be used to offset a portion of this energy.

In Europe, since the mid-1990s about 125 large solar-thermal district heating plants have been constructed, each with over 500 m² (5400 ft²) of solar collectors. The largest are about 10,000 m², with capacities of 7 MW-thermal and solar heat costs around 4 Eurocents/kWh without subsidies. 40 of them have nominal capacities of 1 MW-thermal or more. The Solar District Heating program (SDH) has participation from 14 European Nations and the European Commission, and is working toward technical and market development, and holds annual conferences.

MIT's Solar House #1 built in 1939 used seasonal thermal energy storage (STES) for year-round heating.

Low-temperature Collectors

Glazed solar collectors are designed primarily for space heating. They recirculate building air through a solar air panel where the air is heated and then directed back into the building. These solar space heating systems require at least two penetrations into the building and only perform when the air in the solar collector is warmer than the building room temperature. Most glazed collectors are used in the residential sector.

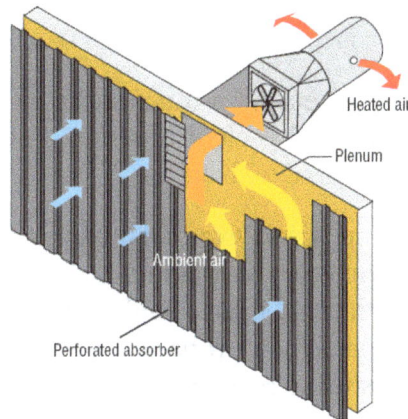

Unglazed, "transpired" air collector

Unglazed solar collectors are primarily used to pre-heat make-up ventilation air in commercial, industrial and institutional buildings with a high ventilation load. They turn building walls or sections of walls into low cost, high performance, unglazed solar collectors. Also called, "transpired solar panels" or "solar wall", they employ a painted perforated metal solar heat absorber that also serves as the exterior wall surface of the building. Heat conducts from the absorber surface to the thermal boundary layer of air 1 mm thick on the outside of the absorber and to air that passes behind the absorber. The boundary layer of air is drawn into a nearby perforation before the heat can escape by convection to the outside air. The heated air is then drawn from behind the absorber plate into the building's ventilation system.

A Trombe wall is a passive solar heating and ventilation system consisting of an air channel sandwiched between a window and a sun-facing thermal mass. During the ventilation cycle, sunlight stores heat in the thermal mass and warms the air channel causing circulation through vents at the top and bottom of the wall. During the heating cycle the Trombe wall radiates stored heat.

Solar roof ponds are unique solar heating and cooling systems developed by Harold Hay in the 1960s. A basic system consists of a roof-mounted water bladder with a movable insulating cover. This system can control heat exchange between interior and exterior environments by covering and uncovering the bladder between night and day. When heating is a concern the bladder is uncovered during the day allowing sunlight to warm the water bladder and store heat for evening use. When cooling is a concern the covered bladder draws heat from the building's interior during the day and is uncovered at night to radiate heat to the cooler atmosphere. The Skytherm house in Atascadero, California uses a prototype roof pond for heating and cooling.

Solar space heating with solar air heat collectors is more popular in the USA and Canada than heating with solar liquid collectors since most buildings already have a ventilation system for heating and cooling. The two main types of solar air panels are glazed and unglazed.

Of the 21,000,000 square feet (2,000,000 m²) of solar thermal collectors produced in the United States in 2007, 16,000,000 square feet (1,500,000 m²) were of the low-temperature variety. Low-temperature collectors are generally installed to heat swimming pools, although they can also be used for space heating. Collectors can use air or water as the medium to transfer the heat to their destination.

Heat Storage in Low-Temperature Solar Thermal Systems

Interseasonal storage. Solar heat (or heat from other sources) can be effectively stored between opposing seasons aquifers, underground geological strata, large specially constructed pits, and large tanks that are insulated and covered with earth.

Short-term storage. Thermal mass materials store solar energy during the day and release this energy during cooler periods. Common thermal mass materials include stone, concrete, and water. The proportion and placement of thermal mass should consider several factors such as climate, daylighting, and shading conditions. When properly incorporated, thermal mass can passively maintain comfortable temperatures while reducing energy consumption.

Solar-Driven Cooling

Worldwide, by 2011 there were about 750 cooling systems with solar-driven heat pumps, and annual market growth was 40 to 70% over the prior seven years. It is a niche market because the economics are challenging, with the annual number of cooling hours a limiting factor. Respectively, the annual cooling hours are roughly 1000 in the Mediterranean, 2500 in Southeast Asia, and only 50 to 200 in Central Europe. However, system construction costs dropped about 50% between 2007 and 2011. The International Energy Agency (IEA) Solar Heating and Cooling program (IEA-SHC) task groups working on further development of the technologies involved.

Solar Heat-driven Ventilation

A solar chimney (or thermal chimney) is a passive solar ventilation system composed of a hollow thermal mass connecting the interior and exterior of a building. As the chimney warms, the air inside is heated causing an updraft that pulls air through the building. These systems have been in use since Roman times and remain common in the Middle East.

Process Heat

Solar process heating systems are designed to provide large quantities of hot water or space heating for nonresidential buildings.

Solar Evaporation Ponds in the Atacama Desert.

Evaporation ponds are shallow ponds that concentrate dissolved solids through evaporation. The use of evaporation ponds to obtain salt from sea water is one of the oldest applications of solar energy. Modern uses include concentrating brine solutions used in leach mining and removing dissolved solids from waste streams. Altogether, evaporation ponds represent one of the largest commercial applications of solar energy in use today.

Unglazed transpired collectors are perforated sun-facing walls used for preheating ventilation air. Transpired collectors can also be roof mounted for year round use and can raise the incoming air temperature up to 22 °C and deliver outlet temperatures of 45-60 °C. The short payback period of transpired collectors (3 to 12 years) make them a more cost-effective alternative to glazed collection systems. As of 2015, over 4000 systems with a combined collector area of 500,000 m² had been installed worldwide. Representatives include an 860 m² collector in Costa Rica used for drying coffee beans and a 1300 m² collector in Coimbatore, India used for drying marigolds.

A food processing facility in Modesto, California uses parabolic troughs to produce steam used in the manufacturing process. The 5,000 m² collector area is expected to provide 15 TJ per year.

Medium-temperature Collectors

These collectors could be used to produce approximately 50% and more of the hot water needed for residential and commercial use in the United States. In the United States, a typical system costs $4000–$6000 retail ($1400 to $2200 wholesale for the materials) and 30% of the system qualifies for a federal tax credit + additional state credit exists in about half of the states. Labor for a simple open loop system in southern climates can take 3–5 hours for the installation and 4–6 hours in Northern areas. Northern system require more collector area and more complex plumbing to protect the collector from freezing. With this incentive, the payback time for a typical household is four to nine years, depending on the state. Similar subsidies exist in parts of Europe. A crew of one solar plumber and two assistants with minimal training can install a system per day. Thermosiphon installation have negligible maintenance costs (costs rise if antifreeze and mains power are used for circulation) and in the US reduces a households' operating costs by $6 per person per month. Solar water heating can reduce CO_2 emissions of a family of four by 1 ton/year (if replacing natural gas) or 3 ton/year (if replacing electricity). Medium-temperature installations can use any of several designs: common designs are pressurized glycol, drain back, batch systems and newer low pressure freeze tolerant systems using polymer pipes containing water with photovoltaic pumping. European and International standards are being reviewed to accommodate innovations in design and operation of medium temperature collectors. Operational innovations include "permanently wetted collector" operation. This innovation reduces or even eliminates the occurrence of no-flow high temperature stresses called stagnation which would otherwise reduce the life expectancy of collectors.

Solar Drying

Solar thermal energy can be useful for drying wood for construction and wood fuels such as wood chips for combustion. Solar is also used for food products such as fruits, grains, and fish. Crop drying by solar means is environmentally friendly as well as cost effective while improving the quality. The less money it takes to make a product, the less it can be sold for, pleasing both the buyers and the sellers. Technologies in solar drying include ultra low cost pumped transpired plate air collectors based on black fabrics. Solar thermal energy is helpful in the process of drying products such as wood chips and other forms of biomass by raising the temperature while allowing air to pass through and get rid of the moisture.

Cooking

Solar cookers use sunlight for cooking, drying and pasteurization. Solar cooking offsets fuel costs, reduces demand for fuel or firewood, and improves air quality by reducing or removing a source of smoke.

Industrial indirect solar fruit and vegetable dryer

The simplest type of solar cooker is the box cooker first built by Horace de Saussure in 1767. A basic box cooker consists of an insulated container with a transparent lid. These cookers can be used effectively with partially overcast skies and will typically reach temperatures of 50–100 °C.

Concentrating solar cookers use reflectors to concentrate solar energy onto a cooking container. The most common reflector geometries are flat plate, disc and parabolic trough type. These designs cook faster and at higher temperatures (up to 350 °C) but require direct light to function properly.

The Solar Bowl above the Solar Kitchen in Auroville, India concentrates sunlight on a movable receiver to produce steam for cooking.

The Solar Kitchen in Auroville, India uses a unique concentrating technology known as the solar bowl. Contrary to conventional tracking reflector/fixed receiver systems, the solar bowl uses a fixed spherical reflector with a receiver which tracks the focus of light as the Sun moves across the sky. The solar bowl's receiver reaches temperature of 150 °C that is used to produce steam that helps cook 2,000 daily meals.

Many other solar kitchens in India use another unique concentrating technology known as the Scheffler reflector. This technology was first developed by Wolfgang Scheffler in 1986. A Scheffler reflector is a parabolic dish that uses single axis tracking to follow the Sun's daily course. These reflectors have a flexible reflective surface that is able to change its curvature to adjust to seasonal variations in the incident angle of sunlight. Scheffler reflectors have the advantage of having a fixed focal point which improves the ease of cooking and are able to reach temperatures of 450-650 °C. Built in 1999 by the Brahma Kumaris, the world's largest Scheffler reflector system in Abu Road, Rajasthan India is capable of cooking up to 35,000 meals a day. By early 2008, over 2000 large cookers of the Scheffler design had been built worldwide.

Distillation

Solar stills can be used to make drinking water in areas where clean water is not common. Solar distillation is necessary in these situations to provide people with purified water. Solar energy heats up the water in the still. The water then evaporates and condenses on the bottom of the covering glass.

High-temperature Collectors

Where temperatures below about 95 °C are sufficient, as for space heating, flat-plate collectors of the nonconcentrating type are generally used. Because of the relatively high heat losses through the glazing, flat plate collectors will not reach temperatures much above 200 °C even when the heat transfer fluid is stagnant. Such temperatures are too low for efficient conversion to electricity.

The efficiency of heat engines increases with the temperature of the heat source. To achieve this in solar thermal energy plants, solar radiation is concentrated by mirrors or lenses to obtain higher temperatures – a technique called Concentrated Solar Power (CSP). The practical effect of high efficiencies is to reduce the plant's collector size and total land use per unit power generated, reducing the environmental impacts of a power plant as well as its expense.

Part of the 354 MW SEGS solar complex in northern San Bernardino County, California.

As the temperature increases, different forms of conversion become practical. Up to 600 °C, steam turbines, standard technology, have an efficiency up to 41%. Above 600 °C, gas turbines can be more efficient. Higher temperatures are problematic because different materials and techniques are needed. One proposal for very high temperatures is to use liquid fluoride salts operating between 700 °C to 800 °C, using multi-stage turbine systems to achieve 50% or more thermal efficiencies. The higher operating temperatures permit the plant to use higher-temperature dry heat exchangers for its thermal exhaust, reducing the plant's water use – critical in the deserts where large solar plants are practical. High temperatures also make heat storage more efficient, because more watt-hours are stored per unit of fluid.

The solar furnace at Odeillo in the French Pyrenees-Orientales can reach temperatures up to 3,500°C .

Commercial concentrating solar thermal power (CSP) plants were first developed in the 1980s. The world's largest solar thermal power plants are now the 370 MW Ivanpah Solar Power Facility, commissioned in 2014, and the 354 MW SEGS CSP installation, both located in the Mojave Desert of California, where several other solar projects have been realized as well. With the exception of the Shams solar power station, built in 2013 near Abu Dhabi, the United Arab Emirates, all other 100 MW or larger CSP plants are either located in the United States or in Spain.

The principal advantage of CSP is the ability to efficiently add thermal storage, allowing the dispatching of electricity over up to a 24-hour period. Since peak electricity demand typically occurs at about 5 pm, many CSP power plants use 3 to 5 hours of thermal storage. With current technology, storage of heat is much cheaper and more efficient than storage of electricity. In this way, the CSP plant can produce electricity day and night. If the CSP site has predictable solar radiation, then the CSP plant becomes a reliable power plant. Reliability can further be improved by installing a back-up combustion system. The back-up system can use most of the CSP plant, which decreases the cost of the back-up system.

CSP facilities utilize high electrical conductivity materials, such as copper, in field power cables, grounding networks, and motors for tracking and pumping fluids, as well as in the main generator and high voltage transformers.

With reliability, unused desert, no pollution, and no fuel costs, the obstacles for large deployment for CSP are cost, aesthetics, land use and similar factors for the necessary connecting high tension lines. Although only a small percentage of the desert is necessary to meet global electricity demand, still a large area must be covered with mirrors or lenses to obtain a significant amount of energy. An important way to decrease cost is the use of a simple design.

When considering land use impacts associated with the exploration and extraction through to transportation and conversion of fossil fuels, which are used for most of our electrical power, utility-scale solar power compares as one of the most land-efficient energy resources available:

The federal government has dedicated nearly 2,000 times more acreage to oil and gas leases than to solar development. In 2010 the Bureau of Land Management approved nine large-scale solar projects, with a total generating capacity of 3,682 megawatts, representing approximately 40,000 acres. In contrast, in 2010, the Bureau of Land Management processed more than 5,200 applications gas and oil leases, and issued 1,308 leases, for a total of 3.2 million acres. Currently, 38.2 million acres of onshore public lands and an additional 36.9 million acres of offshore exploration in the Gulf of Mexico are under lease for oil and gas development, exploration and production.

System Designs

During the day the sun has different positions. For low concentration systems (and low temperatures) tracking can be avoided (or limited to a few positions per year) if nonimaging optics are used. For higher concentrations, however, if the mirrors or lenses do not move, then the focus of the mirrors or lenses changes (but also in these cases nonimaging optics provides the widest acceptance angles for a given concentration). Therefore, it seems unavoidable that there needs to be a tracking system that follows the position of the sun (for solar photovoltaic a solar tracker is only optional). The tracking system increases the cost and complexity. With this in mind, different designs can be distinguished in how they concentrate the light and track the position of the sun.

Parabolic Trough Designs

Parabolic trough power plants use a curved, mirrored trough which reflects the direct solar radiation onto a glass tube containing a fluid (also called a receiver, absorber or collector) running the length of the trough, positioned at the focal point of the reflectors. The trough is parabolic along one axis and linear in the orthogonal axis. For change of the daily position of the sun perpendicular to the receiver, the trough tilts east to west so that the direct radiation remains focused on the receiver. However, seasonal changes in the angle of sunlight parallel to the trough does not require adjustment of the mirrors, since the light is simply concentrated elsewhere on the receiver. Thus the trough design does not require tracking on a second axis. The receiver may be enclosed in a glass vacuum chamber. The vacuum significantly reduces convective heat loss.

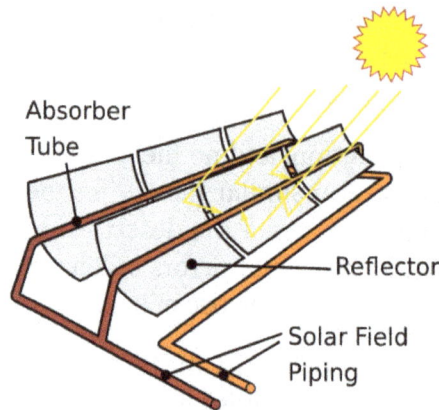

Sketch of a parabolic trough design. A change of position of the sun parallel to the receiver does not require adjustment of the mirrors.

A fluid (also called heat transfer fluid) passes through the receiver and becomes very hot. Common fluids are synthetic oil, molten salt and pressurized steam. The fluid containing the heat is transported to a heat engine where about a third of the heat is converted to electricity.

Full-scale parabolic trough systems consist of many such troughs laid out in parallel over a large area of land. Since 1985 a solar thermal system using this principle has been in full operation in California in the United States. It is called the Solar Energy Generating Systems (SEGS) system. Other CSP designs lack this kind of long experience and therefore it can currently be said that the parabolic trough design is the most thoroughly proven CSP technology.

The SEGS is a collection of nine plants with a total capacity of 354 MW and has been the world's largest solar power plant, both thermal and non-thermal, for many years. A newer plant is Nevada Solar One plant with a capacity of 64 MW. The 150 MW Andasol solar power stations are in Spain with each site having a capacity of 50 MW. Note however, that those plants have heat storage which requires a larger field of solar collectors relative to the size of the steam turbine-generator to store heat and send heat to the steam turbine at the same time. Heat storage enables better utilization of the steam turbine. With day and some nighttime operation of the steam-turbine Andasol 1 at 50 MW peak capacity produces more energy than Nevada Solar One at 64 MW peak capacity, due to the former plant's thermal energy storage system and larger solar field. The 280MW Solana Generating Station came online in Arizona in 2013 with 6 hours of power storage. Hassi R'Mel integrated solar combined cycle power station in Algeria and Martin Next Generation Solar Energy Center both use parabolic troughs in a combined cycle with natural gas.

Enclosed Trough

The enclosed trough architecture encapsulates the solar thermal system within a greenhouse-like glasshouse. The glasshouse creates a protected environment to withstand the elements that can negatively impact reliability and efficiency of the solar thermal system.

Inside an enclosed trough system

Lightweight curved solar-reflecting mirrors are suspended within the glasshouse structure. A single-axis tracking system positions the mirrors to track the sun and focus its light onto a network of stationary steel pipes, also suspended from the glasshouse structure. Steam is generated directly, using oil field-quality water, as water flows from the inlet throughout the length of the pipes, without heat exchangers or intermediate working fluids.

The steam produced is then fed directly to the field's existing steam distribution network, where the steam is continuously injected deep into the oil reservoir. Sheltering the mirrors from the wind allows them to achieve higher temperature rates and prevents dust from building up as a result from exposure to humidity. GlassPoint Solar, the company that created the Enclosed Trough design, states its technology can produce heat for EOR for about $5 per million British thermal units in sunny regions, compared to between $10 and $12 for other conventional solar thermal technologies.

Power Tower Designs

Ivanpah Solar Electric Generating System with all three towers under load, Feb., 2014. Taken from I-15 in San Bernardino County, California. The Clark Mountain Range can be seen in the distance.

Power towers (also known as 'central tower' power plants or 'heliostat' power plants) capture and focus the sun's thermal energy with thousands of tracking mirrors (called heliostats) in roughly a two square mile field. A tower resides in the center of the heliostat field. The heliostats focus concentrated sunlight on a receiver which sits on top of the tower. Within the receiver the concentrated sunlight heats molten salt to over 1,000 °F (538 °C). The heated molten salt then flows into a thermal storage tank where it is stored, maintaining 98% thermal efficiency, and eventually pumped to a steam generator. The steam drives a standard turbine to generate electricity. This process, also known as the "Rankine cycle" is similar to a standard coal-fired power plant, except it is fueled by clean and free solar energy.

The advantage of this design above the parabolic trough design is the higher temperature. Thermal energy at higher temperatures can be converted to electricity more efficiently and can be more cheaply stored for later use. Furthermore, there is less need to flatten the ground area. In principle a power tower can be built on the side of a hill. Mirrors can be flat and plumbing is concentrated in the tower. The disadvantage is that each mirror must have its own dual-axis control, while in the parabolic trough design single axis tracking can be shared for a large array of mirrors.

A cost/performance comparison between power tower and parabolic trough concentrators was made by the NREL which estimated that by 2020 electricity could be produced from power towers for 5.47 ¢/kWh and for 6.21 ¢/kWh from parabolic troughs. The capacity factor for power towers was estimated to be 72.9% and 56.2% for parabolic troughs. There is some hope that the development of cheap, durable, mass producible heliostat power plant components could bring this cost down.

The first commercial tower power plant was PS10 in Spain with a capacity of 11 MW, completed in 2007. Since then a number of plants have been proposed, several have been built on a number of countries (Spain, Germany, U.S., Turkey, China, India) but several proposed plants were cancelled as photovoltaic solar prices plummeted. A solar power tower is expected to come online in South Africa in 2014. Ivanpah Solar Power Facility in California generates 392 MW of electricity from three towers, making it the largest solar power tower plant when it came online in late 2013.

Dish Designs

A parabolic solar dish concentrating the sun's rays on the heating element of a Stirling engine. The entire unit acts as a solar tracker.

CSP-Stirling is known to have the highest efficiency of all solar technologies (around 30%, compared to solar photovoltaic's approximately 15%), and is predicted to be able to produce the cheapest energy among all renewable energy sources in high-scale production and hot areas, semi-deserts, etc. A dish Stirling system uses a large, reflective, parabolic dish (similar in shape to a satellite television dish). It focuses all the sunlight that strikes the dish up onto a single point above the dish, where a receiver captures the heat and transforms it into a useful form. Typically the dish is coupled with a Stirling engine in a Dish-Stirling System, but also sometimes a steam engine is used. These create rotational kinetic energy that can be converted to electricity using an electric generator.

In 2005 Southern California Edison announced an agreement to purchase solar powered Stirling engines from Stirling Energy Systems over a twenty-year period and in quantities (20,000 units) sufficient to generate 500 megawatts of electricity. In January 2010, Stirling Energy Systems and Tessera Solar commissioned the first demonstration 1.5-megawatt power plant ("Maricopa Solar") using Stirling technology in Peoria, Arizona. At the beginning of 2011 Stirling Energy's development arm, Tessera Solar, sold off its two large projects, the 709 MW Imperial project and the 850 MW Calico project to AES Solar and K.Road, respectively. In 2012 the Maricopa plant was bought and dismantled by United Sun Systems. United Sun Systems released a new generation system, based on a V-shaped Stirling engine and a peak production of 33 kW. The new CSP-Stirling technology brings down LCOE to USD 0.02 in utility scale.

According to its developer, Rispasso Energy, a Swedish firm, in 2015 its Dish Sterling system being tested in the Kalahari Desert in South Africa showed 34% efficiency.

Fresnel Technologies

A linear Fresnel reflector power plant uses a series of long, narrow, shallow-curvature (or even flat) mirrors to focus light onto one or more linear receivers positioned above the mirrors. On top of the receiver a small parabolic mirror can be attached for further focusing the light. These systems aim to offer lower overall costs by sharing a receiver between several mirrors (as compared with trough and dish concepts), while still using the simple line-focus geometry with one axis for tracking. This is similar to the trough design (and different from central towers and dishes with dual-axis). The receiver is stationary and so fluid couplings are not required (as in troughs and dishes). The mirrors also do not need to support the receiver, so they are structurally simpler. When suitable aiming strategies are used (mirrors aimed at different receivers at different times of day), this can allow a denser packing of mirrors on available land area.

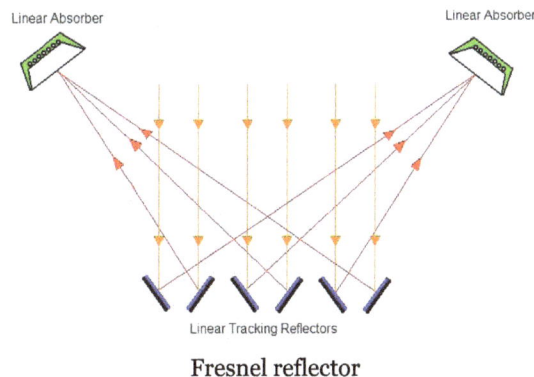

Fresnel reflector

Rival single axis tracking technologies include the relatively new linear Fresnel reflector (LFR) and compact-LFR (CLFR) technologies. The LFR differs from that of the parabolic trough in that the absorber is fixed in space above the mirror field. Also, the reflector is composed of many low row segments, which focus collectively on an elevated long tower receiver running parallel to the reflector rotational axis.

Prototypes of Fresnel lens concentrators have been produced for the collection of thermal energy by International Automated Systems. No full-scale thermal systems using Fresnel lenses are known to be in operation, although products incorporating Fresnel lenses in conjunction with photovoltaic cells are already available.

MicroCSP

MicroCSP is used for community-sized power plants (1 MW to 50 MW), for industrial, agricultural and manufacturing 'process heat' applications, and when large amounts of hot water are needed, such as resort swimming pools, water parks, large laundry facilities, sterilization, distillation and other such uses.

Enclosed Parabolic Trough

The enclosed parabolic trough solar thermal system encapsulates the components within an off-the-shelf greenhouse type of glasshouse. The glasshouse protects the components from the elements that can negatively impact system reliability and efficiency. This protection importantly includes nightly glass-roof washing with optimized water-efficient off-the-shelf automated washing systems. Lightweight curved solar-reflecting mirrors are suspended from the ceiling of the glasshouse by wires. A single-axis tracking system positions the mirrors to retrieve the optimal amount of sunlight. The mirrors concentrate the sunlight and focus it on a network of stationary steel pipes, also suspended from the glasshouse structure. Water is pumped through the pipes and boiled to generate steam when intense sun radiation is applied. The steam is available for process heat. Sheltering the mirrors from the wind allows them to achieve higher temperature rates and prevents dust from building up on the mirrors as a result from exposure to humidity.

Heat Collection and Exchange

More energy is contained in higher frequency light based upon the formula of , where h is the Planck constant and is frequency. Metal collectors down convert higher frequency light by producing a series of Compton shifts into an abundance of lower frequency light. Glass or ceramic coatings with high transmission in the visible and UV and effective absorption in the IR (heat blocking) trap metal absorbed low frequency light from radiation loss. Convection insulation prevents mechanical losses transferred through gas. Once collected as heat, thermos containment efficiency improves significantly with increased size. Unlike Photovoltaic technologies that often degrade under concentrated light, Solar Thermal depends upon light concentration that requires a clear sky to reach suitable temperatures.

Heat in a solar thermal system is guided by five basic principles: heat gain; heat transfer; heat storage; heat transport; and heat insulation. Here, heat is the measure of the amount of thermal energy an object contains and is determined by the temperature, mass and specific heat of the object. Solar thermal power plants use heat exchangers that are designed for constant working conditions, to provide heat exchange. Copper heat exchangers are important in solar thermal heating and cooling systems because of copper's high thermal conductivity, resistance to atmospheric and water corrosion, sealing and joining by soldering, and mechanical strength. Copper is used both in receivers and in primary circuits (pipes and heat exchangers for water tanks) of solar thermal water systems.

Heat gain is the heat accumulated from the sun in the system. Solar thermal heat is trapped using the greenhouse effect; the greenhouse effect in this case is the ability of a reflective surface to transmit short wave radiation and reflect long wave radiation. Heat and infrared radiation (IR) are produced when short wave radiation light hits the absorber plate, which is then trapped inside the collector. Fluid, usually water, in the absorber tubes collect the trapped heat and transfer it to a heat storage vault.

Heat is transferred either by conduction or convection. When water is heated, kinetic energy is transferred by conduction to water molecules throughout the medium. These molecules spread their thermal energy by conduction and occupy more space than the cold slow moving molecules above them. The distribution of energy from the rising hot water to the sinking cold water contributes to the convection process. Heat is transferred from the absorber plates of the collector in the fluid by conduction. The collector fluid is circulated through the carrier pipes to the heat transfer vault. Inside the vault, heat is transferred throughout the medium through convection.

Heat storage enables solar thermal plants to produce electricity during hours without sunlight. Heat is transferred to a thermal storage medium in an insulated reservoir during hours with sunlight, and is withdrawn for power generation during hours lacking sunlight. Thermal storage mediums will be discussed in a heat storage section. Rate of heat transfer is related to the conductive and convection medium as well as the temperature differences. Bodies with large temperature differences transfer heat faster than bodies with lower temperature differences.

Heat transport refers to the activity in which heat from a solar collector is transported to the heat storage vault. Heat insulation is vital in both heat transport tubing as well as the storage vault. It prevents heat loss, which in turn relates to energy loss, or decrease in the efficiency of the system.

Heat Storage for Space Heating

A collection of mature technologies called seasonal thermal energy storage (STES) is capable of storing heat for months at a time, so solar heat collected primarily in Summer can be used for all-year heating. Solar-supplied STES technology has been advanced primarily in Denmark, Germany, and Canada, and applications include individual buildings and district heating networks. Drake Landing Solar Community in Alberta, Canada has a small district system and in 2012 achieved a world record of providing 97% of the a community's all-year space heating needs from the sun. STES thermal storage mediums include deep aquifers; native rock surrounding clusters of small-diameter, heat exchanger equipped boreholes; large, shallow, lined pits that are filled with gravel and top-insulated; and large, insulated and buried surface water tanks.

Heat Storage to Stabilize Solar-electric Power Generation

Heat storage allows a solar thermal plant to produce electricity at night and on overcast days. This allows the use of solar power for baseload generation as well as peak power generation, with the potential of displacing both coal- and natural gas-fired power plants. Additionally, the utilization of the generator is higher which reduces cost.

Heat is transferred to a thermal storage medium in an insulated reservoir during the day, and withdrawn for power generation at night. Thermal storage media include pressurized steam, concrete, a variety of phase change materials, and molten salts such as calcium, sodium and potassium nitrate.

Steam Accumulator

The PS10 solar power tower stores heat in tanks as pressurized steam at 50 bar and 285 °C. The steam condenses and flashes back to steam, when pressure is lowered. Storage is for one hour. It is suggested that longer storage is possible, but that has not been proven yet in an existing power plant.

Molten Salt Storage

A variety of fluids have been tested to transport the sun's heat, including water, air, oil, and sodium, but Rockwell International selected molten salt as best. Molten salt is used in solar power tower systems because it is liquid at atmospheric pressure, provides a low-cost medium to store thermal energy, its operating temperatures are compatible with today's steam turbines, and it is non-flammable and nontoxic. Molten salt is used in the chemical and metals industries to transport heat, so industry has experience with it.

The 150 MW Andasol solar power station is a commercial parabolic trough solar thermal power plant, located in Spain. The Andasol plant uses tanks of molten salt to store solar energy so that it can continue generating electricity even when the sun isn't shining.

The first commercial molten salt mixture was a common form of saltpeter, 60% sodium nitrate and 40% potassium nitrate. Saltpeter melts at 220 °C (430 °F) and is kept liquid at 290 °C (550 °F) in an insulated storage tank. Calcium nitrate can reduce the melting point to 131 °C, permitting more energy to be extracted before the salt freezes. There are now several technical calcium nitrate grades stable at more than 500 °C.

This solar power system can generate power in cloudy weather or at night using the heat in the tank of hot salt. The tanks are insulated, able to store heat for a week. Tanks that power a 100-megawatt turbine for four hours would be about 9 m (30 ft) tall and 24 m (80 ft) in diameter.

The Andasol power plant in Spain is the first commercial solar thermal power plant using molten salt for heat storage and nighttime generation. It came on line March 2009. On July 4, 2011, a company in Spain celebrated an historic moment for the solar industry: Torresol's 19.9 MW concentrating solar power plant became the first ever to generate uninterrupted electricity for 24 hours straight, using a molten salt heat storage.

Phase-change Materials for Storage

Phase Change Material (PCMs) offer an alternative solution in energy storage. Using a similar heat transfer infrastructure, PCMs have the potential of providing a more efficient means of storage. PCMs can be either organic or inorganic materials. Advantages of organic PCMs include no corrosives, low or no undercooling, and chemical and thermal stability. Disadvantages include low phase-change enthalpy, low thermal conductivity, and flammability. Inorganics are advantageous with greater phase-change enthalpy, but exhibit disadvantages with undercooling, corrosion, phase separation, and lack of thermal stability. The greater phase-change enthalpy in inorganic PCMs make hydrate salts a strong candidate in the solar energy storage field.

Use of Water

A design which requires water for condensation or cooling may conflict with location of solar thermal plants in desert areas with good solar radiation but limited water resources. The conflict is illustrated by plans of Solar Millennium, a German company, to build a plant in the Amargosa Valley of Nevada which would require 20% of the water available in the area. Some other projected plants by the same and other companies in the Mojave Desert of California may also be affected by difficulty in obtaining adequate and appropriate water rights. California water law currently prohibits use of potable water for cooling.

Other designs require less water. The Ivanpah Solar Power Facility in south-eastern California conserves scarce desert water by using air-cooling to convert the steam back into water. Compared to conventional wet-cooling, this results in a 90% reduction in water usage at the cost of some loss of efficiency. The water is then returned to the boiler in a closed process which is environmentally friendly.

Conversion Rates from Solar Energy to Electrical Energy

Of all of these technologies the solar dish/Stirling engine has the highest energy efficiency. A single solar dish-Stirling engine installed at Sandia National Laboratories National Solar Thermal Test Facility (NSTTF) produces as much as 25 kW of electricity, with a conversion efficiency of 31.25%.

Solar parabolic trough plants have been built with efficiencies of about 20%. Fresnel reflectors have an efficiency that is slightly lower (but this is compensated by the denser packing).

The gross conversion efficiencies (taking into account that the solar dishes or troughs occupy only a fraction of the total area of the power plant) are determined by net generating capacity over the solar energy that falls on the total area of the solar plant. The 500-megawatt (MW)

SCE/SES plant would extract about 2.75% of the radiation (1 kW/m²; see Solar power) that falls on its 4,500 acres (18.2 km²). For the 50 MW AndaSol Power Plant that is being built in Spain (total area of 1,300×1,500 m = 1.95 km²) gross conversion efficiency comes out at 2.6%.

Furthermore, efficiency does not directly relate to cost: on calculating total cost, both efficiency and the cost of construction and maintenance should be taken into account.

Solar Power

Average insolation. Note that this is for a horizontal surface, whereas solar panels are normally propped up at an angle and receive more energy per unit area, especially at high latitudes. Potential of solar energy. The small black dots show land area required to replace the world primary energy supply with solar power.

A solar PV array on a rooftop in Hong Kong

The first three concentrated solar power (CSP) units of Spain's Solnova Solar Power Station in the foreground, with the PS10 and PS20 solar power towers in the background

The International Energy Agency projected in 2014 that under its "high renewables" scenario, by 2050, solar photovoltaics and concentrated solar power would contribute about 16 and 11 percent, respectively, of the worldwide electricity consumption, and solar would be the world's largest source of electricity. Most solar installations would be in China and India.

Photovoltaics were initially solely used as a source of electricity for small and medium-sized applications, from the calculator powered by a single solar cell to remote homes powered by an off-grid rooftop PV system. As the cost of solar electricity has fallen, the number of grid-connected solar PV systems has grown into the millions and utility-scale solar power stations with hundreds of megawatts are being built. Solar PV is rapidly becoming an inexpensive, low-carbon technology to harness renewable energy from the Sun.

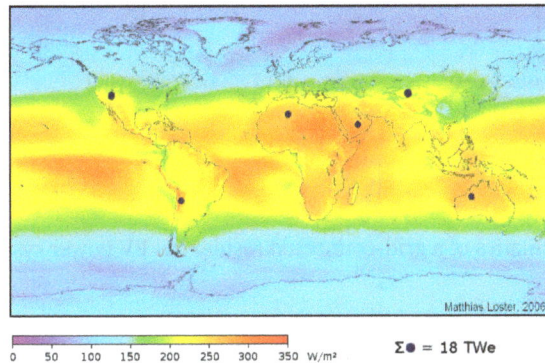

Solar power is the conversion of sunlight into electricity, either directly using photovoltaics (PV), or indirectly using concentrated solar power. Concentrated solar power systems (Unified Solar) use lenses or mirrors and tracking systems to focus a large area of sunlight into a small beam. Photovoltaics convertlight into an electric current using the photovoltaic effect.

Commercial concentrated solar power plants were first developed in the 1980s. The 392 MW Ivanpah installation is the largest concentrating solar power plant in the world, located in the Mojave Desert of California.

Mainstream Technologies

Many industrialized nations have installed significant solar power capacity into their grids to supplement or provide an alternative to conventional energy sources while an increasing number of less developed nations have turned to solar to reduce dependence on expensive imported fuels. Long distance transmission allows remote renewable energy resources to displace fossil fuel consumption. Solar power plants use one of two technologies:

- Photovoltaic (PV) systems use solar panels, either on rooftops or in ground-mounted solar farms, converting sunlight directly into electric power.

- Concentrated solar power (CSP, also known as "concentrated solar thermal") plants use solar thermal energy to make steam, that is thereafter converted into electricity by a turbine.

Photovoltaics

A solar cell, or photovoltaic cell (PV), is a device that converts light into electric current using the photovoltaic effect. The first solar cell was constructed by Charles Fritts in the 1880s. The German industrialist Ernst Werner von Siemens was among those who recognized the importance of this discovery. In 1931, the German engineer Bruno Lange developed a photo cell using silver selenide in place of copper oxide, although the prototype selenium cells converted less than 1% of incident light into electricity. Following the work of Russell Ohl in the 1940s, researchers Gerald Pearson, Calvin Fuller and Daryl Chapin created the silicon solar cell in 1954. These early solar cells cost 286 USD/watt and reached efficiencies of 4.5–6%.

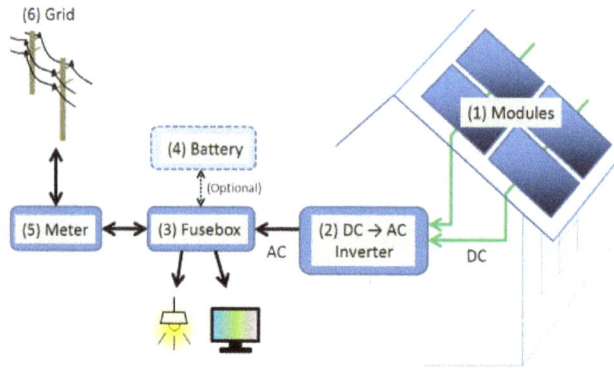

Schematics of a grid-connected residential PV power system

Conventional PV Systems

The array of a photovoltaic power system, or PV system, produces direct current (DC) power which fluctuates with the sunlight's intensity. For practical use this usually requires conversion to certain desired voltages or alternating current (AC), through the use of inverters. Multiple solar cells are connected inside modules. Modules are wired together to form arrays, then tied to an inverter, which produces power at the desired voltage, and for AC, the desired frequency/phase.

Many residential PV systems are connected to the grid wherever available, especially in developed countries with large markets. In these grid-connected PV systems, use of energy storage is optional. In certain applications such as satellites, lighthouses, or in developing countries, batteries or additional power generators are often added as back-ups. Such stand-alone power systems permit operations at night and at other times of limited sunlight.

Concentrated Solar Power

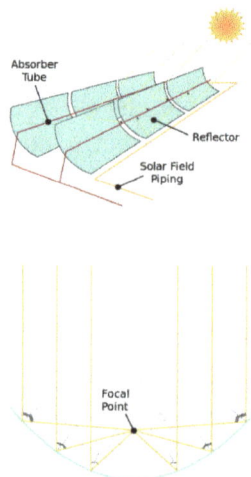

A parabolic collector concentrates sunlight onto a tube in its focal point.

Concentrated solar power (CSP), also called "concentrated solar thermal", uses lenses or mirrors and tracking systems to focus a large area of sunlight into a small beam. Contrary to photovoltaics – which converts light directly into electricity – CSP uses the heat of the sun's radiation to generate electricity from conventional steam-driven turbines.

A wide range of concentrating technologies exists: among the best known are the parabolic trough, the compact linear Fresnel reflector, the Stirling dish and the solar power tower. Various techniques are used to track the sun and focus light. In all of these systems a working fluid is heated by the concentrated sunlight, and is then used for power generation or energy storage. Thermal storage efficiently allows up to 24-hour electricity generation.

A *parabolic trough* consists of a linear parabolic reflector that concentrates light onto a receiver positioned along the reflector's focal line. The receiver is a tube positioned right above the middle of the parabolic mirror and is filled with a working fluid. The reflector is made to follow the sun during daylight hours by tracking along a single axis. Parabolic trough systems provide the best land-use factor of any solar technology. The SEGS plants in California and Acciona's Nevada Solar One near Boulder City, Nevada are representatives of this technology.

Compact Linear Fresnel Reflectors are CSP-plants which use many thin mirror strips instead of parabolic mirrors to concentrate sunlight onto two tubes with working fluid. This has the advantage that flat mirrors can be used which are much cheaper than parabolic mirrors, and that more reflectors can be placed in the same amount of space, allowing more of the available sunlight to be used. Concentrating linear fresnel reflectors can be used in either large or more compact plants.

The *Stirling solar dish* combines a parabolic concentrating dish with a Stirling engine which normally drives an electric generator. The advantages of Stirling solar over photovoltaic cells are higher efficiency of converting sunlight into electricity and longer lifetime. Parabolic dish systems give the highest efficiency among CSP technologies. The 50 kW Big Dish in Canberra, Australia is an example of this technology.

A *solar power tower* uses an array of tracking reflectors (heliostats) to concentrate light on a central receiver atop a tower. Power towers are more cost effective, offer higher efficiency and better energy storage capability among CSP technologies. The PS10 Solar Power Plant and PS20 solar power plant are examples of this technology.

Hybrid Systems

A hybrid system combines (C)PV and CSP with one another or with other forms of generation such as diesel, wind and biogas. The combined form of generation may enable the system to modulate power output as a function of demand or at least reduce the fluctuating nature of solar power and the consumption of non renewable fuel. Hybrid systems are most often found on islands.

CPV/CSP system

A novel solar CPV/CSP hybrid system has been proposed, combining concentrator photovoltaics with the non-PV technology of concentrated solar power, or also known as concentrated solar thermal.

ISCC system

The Hassi R'Mel power station in Algeria, is an example of combining CSP with a gas turbine, where a 25-megawatt CSP-parabolic trough array supplements a much larger 130 MW combined cycle gas turbine plant. Another example is the Yazd power station in Iran.

PVT system

Hybrid PV/T), also known as *photovoltaic thermal hybrid solar collectors* convert solar radiation into thermal and electrical energy. Such a system combines a solar (PV) module with a solar thermal collector in an complementary way.

CPVT system

A concentrated photovoltaic thermal hybrid (CPVT) system is similar to a PVT system. It uses concentrated photovoltaics (CPV) instead of conventional PV technology, and combines it with a solar thermal collector.

PV diesel system

It combines a photovoltaic system with a diesel generator. Combinations with other renewables are possible and include wind turbines.

PV-thermoelectric system

Thermoelectric, or "thermovoltaic" devices convert a temperature difference between dissimilar materials into an electric current. Solar cells use only the high frequency part of the radiation, while the low frequency heat energy is wasted. Several patents about the use of thermoelectric devices in tandem with solar cells have been filed. The idea is to increase the efficiency of the combined solar/thermoelectric system to convert the solar radiation into useful electricity.

Development and Deployment

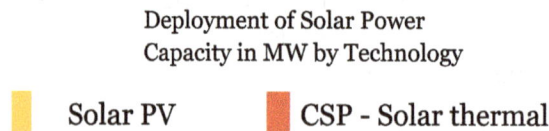

Deployment of Solar Power
Capacity in MW by Technology

Solar PV   CSP - Solar thermal

Electricity Generation from Solar		
Year	Energy (TWh)	% of Total
2004	2.6	0.01%
2005	3.7	0.02%
2006	5.0	0.03%
2007	6.8	0.03%
2008	11.4	0.06%
2009	19.3	0.10%

2010	31.4	0.15%
2011	60.6	0.27%
2012	96.7	0.43%
2013	134.5	0.58%
2014	185.9	0.79%
2015	253.0	1.05%
Source: BP-Statistical Review of World Energy, 2016		

Early Days

The early development of solar technologies starting in the 1860s was driven by an expectation that coal would soon become scarce. However, development of solar technologies stagnated in the early 20th century in the face of the increasing availability, economy, and utility of coal and petroleum. In 1974 it was estimated that only six private homes in all of North America were entirely heated or cooled by functional solar power systems. The 1973 oil embargo and 1979 energy crisis caused a re-organization of energy policies around the world and brought renewed attention to developing solar technologies. Deployment strategies focused on incentive programs such as the Federal Photovoltaic Utilization Program in the US and the Sunshine Program in Japan. Other efforts included the formation of research facilities in the United States (SERI, now NREL), Japan (NEDO), and Germany (Fraunhofer–ISE). Between 1970 and 1983 installations of photovoltaic systems grew rapidly, but falling oil prices in the early 1980s moderated the growth of photovoltaics from 1984 to 1996.

Mid-1990s to Early 2010s

In the mid-1990s, development of both, residential and commercial rooftop solar as well as utility-scale photovoltaic power stations, began to accelerate again due to supply issues with oil and natural gas, global warming concerns, and the improving economic position of PV relative to other energy technologies. In the early 2000s, the adoption of feed-in tariffs—a policy mechanism, that gives renewables priority on the grid and defines a fixed price for the generated electricity—lead to a high level of investment security and to a soaring number of PV deployments in Europe.

Current Status

For several years, worldwide growth of solar PV was driven by European deployment, but has since shifted to Asia, especially China and Japan, and to a growing number of countries and regions all over the world, including, but not limited to, Australia, Canada, Chile, India, Israel, Mexico, South Africa, South Korea, Thailand, and the United States.

Worldwide growth of photovoltaics has averaged 40% per year since 2000 and total installed capacity reached 139 GW at the end of 2013 with Germany having the most cumulative installations (35.7 GW) and Italy having the highest percentage of electricity generated by solar PV (7.0%).

Concentrated solar power (CSP) also started to grow rapidly, increasing its capacity nearly tenfold from 2004 to 2013, albeit from a lower level and involving fewer countries than solar PV. As of the end of 2013, worldwide cumulative CSP-capacity reached 3,425 MW.

Forecasts

In 2010, the International Energy Agency predicted that global solar PV capacity could reach 3,000 GW or 11% of projected global electricity generation by 2050—enough to generate 4,500 TWh of electricity. Four years later, in 2014, the agency projected that, under its "high renewables" scenario, solar power could supply 27% of global electricity generation by 2050 (16% from PV and 11% from CSP). In 2015, analysts predicted that one million homes in the U.S. will have solar power by the end of 2016.

Photovoltaic Power Stations

The Desert Sunlight Solar Farm is a 550 MW power plant in Riverside County, California, that uses thin-film CdTe-modules made by First Solar. As of November 2014, the 550 megawatt Topaz Solar Farm was the largest photovoltaic power plant in the world. This was surpassed by the 579 MW Solar Star complex. The current largest photovoltaic power station in the world is Longyangxia Dam Solar Park, in Gonghe County, Qinghai, China.

World's largest photovoltaic power stations as of 2015			
Name	Capacity (MW)	Location	Year Completed Info
Longyangxia Dam Solar Park	850	Qinghai, China	2013, 2015
Solar Star I and II	579	California, USA	2015
Topaz Solar Farm	550	California, USA	2014
Desert Sunlight Solar Farm	550	California, USA	2015
California Valley Solar Ranch	292	California, USA	2013
Agua Caliente Solar Project	290	Arizona, USA	2014
Mount Signal Solar	266	California, USA	2014
Antelope Valley Solar Ranch	266	California, USA	*pending*
Charanka Solar Park	224	Gujarat, India	2012
Mesquite Solar project	207	Arizona, USA	*pending (planned 700 MW)*
Huanghe Hydropower Golmud Solar Park	200	Qinghai, China	2011
Gonghe Industrial Park Phase I	200	China	2013
Imperial Valley Solar Project	200	California, USA	2013
Note: figures rounded. List may change frequently.			

Concentrating Solar Power Stations

Commercial concentrating solar power (CSP) plants, also called "solar thermal power stations", were first developed in the 1980s. The 377 MW Ivanpah Solar Power Facility, located in California's Mojave Desert, is the world's largest solar thermal power plant project. Other large CSP plants include the Solnova Solar Power Station (150 MW), the Andasol solar power station (150 MW), and Extresol Solar Power Station (150 MW), all in Spain. The principal advantage of CSP is the ability to efficiently add thermal storage, allowing the dispatching of electricity over up to a 24-hour period. Since peak electricity demand typically occurs at about 5 pm, many CSP power plants use 3 to 5 hours of thermal storage.

Largest operational solar thermal power stations			
Name	**Capacity (MW)**	**Location**	**Notes**
Ivanpah Solar Power Facility	392	Mojave Desert, California, USA	Operational since February 2014. Located southwest of Las Vegas.
Solar Energy Generating Systems	354	Mojave Desert, California, USA	Commissioned between 1984 and 1991. Collection of 9 units.
Mojave Solar Project	280	Barstow, California, USA	Completed December 2014
Solana Generating Station	280	Gila Bend, Arizona, USA	Completed October 2013 Includes a 6h thermal energy storage
Genesis Solar Energy Project	250	Blythe, California, USA	Completed April 2014
Solaben Solar Power Station	200	Logrosán, Spain	Completed 2012–2013
Noor I	160	Morocco	Completed 2016
Solnova Solar Power Station	150	Seville, Spain	Completed in 2010
Andasol solar power station	150	Granada, Spain	Completed 2011. Includes a 7.5h thermal energy storage.
Extresol Solar Power Station	150	Torre de Miguel Sesmero, Spain	Completed 2010–2012 Extresol 3 includes a 7.5h thermal energy storage

Ivanpah Solar Electric Generating System with all three towers under load during February 2014, with the Clark Mountain Range seen in the distance

Part of the 354 MW Solar Energy Generating Systems (SEGS) parabolic trough solar complex in northern San Bernardino County, California

Economics

Cost

Adjusting for inflation, it cost $96 per watt for a solar module in the mid-1970s. Process improvements and a very large boost in production have brought that figure down to 68 cents per watt in February 2016, according to data from Bloomberg New Energy Finance. Palo Alto California signed a wholesale purchase agreement in 2016 that secured solar power for 3.7 cents per kilowatt-hour. And in sunny Dubai large-scale solar generated electricity sold in 2016 for just 2.99 cents per kilowatt-hour -- "competitive with any form of fossil-based electricity — and cheaper than most."

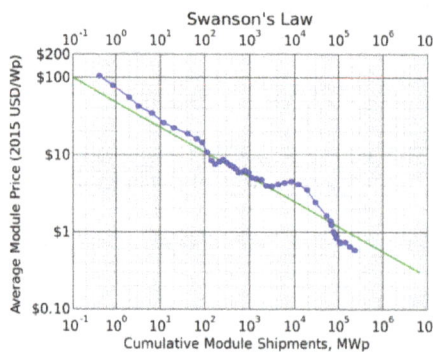

Swanson's law – the PV learning curve

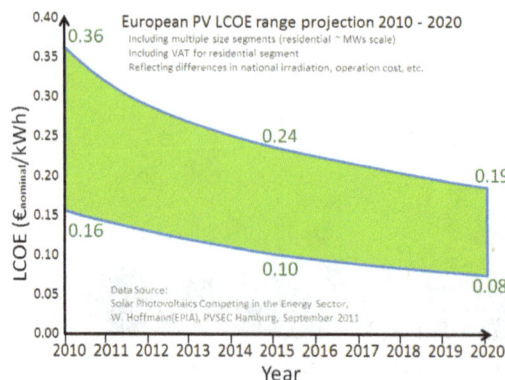

Solar PV – LCOE for Europe until 2020 (in euro-cts. per kWh)

Photovoltaic systems use no fuel, and modules typically last 25 to 40 years. Thus, capital costs make up most of the cost of solar power. Operations and maintenance costs for new utility-scale solar plants in the US are estimated to be 9 percent of the cost of photovoltaic electricity, and 17 percent of the cost of solar thermal electricity. Governments have created various financial incentives to encourage the use of solar power, such as feed-in tariff programs. Also, Renewable portfolio standards impose a government mandate that utilities generate or acquire a certain percentage of renewable power regardless of increased energy procurement costs. In most states, RPS goals can be achieved by any combination of solar, wind, biomass, landfill gas, ocean, geothermal, municipal solid waste, hydroelectric, hydrogen, or fuel cell technologies.

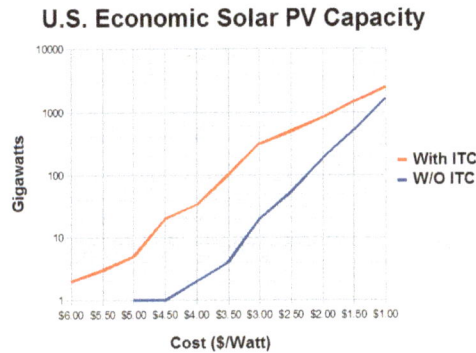

Economic photovoltaic capacity vs installation cost, in the United States

Levelized Cost of Electricity

The PV industry is beginning to adopt levelized cost of electricity (LCOE) as the unit of cost. The electrical energy generated is sold in units of kilowatt-hours (kWh). As a rule of thumb, and depending on the local insolation, 1 watt-peak of installed solar PV capacity generates about 1 to 2 kWh of electricity per year. This corresponds to a capacity factor of around 10–20%. The product of the local cost of electricity and the insolation determines the break even point for solar power. The International Conference on Solar Photovoltaic Investments, organized by EPIA, has estimated that PV systems will pay back their investors in 8 to 12 years. As a result, since 2006 it has been economical for investors to install photovoltaics for free in return for a long term power purchase agreement. Fifty percent of commercial systems in the United States were installed in this manner in 2007 and over 90% by 2009.

Shi Zhengrong has said that, as of 2012, unsubsidised solar power is already competitive with fossil fuels in India, Hawaii, Italy and Spain. He said "We are at a tipping point. No longer are renewable power sources like solar and wind a luxury of the rich. They are now starting to compete in the real world without subsidies". "Solar power will be able to compete without subsidies against conventional power sources in half the world by 2015".

Current Installation Prices

In its 2014 edition of the *Technology Roadmap: Solar Photovoltaic Energy* report, the International Energy Agency (IEA) published prices for residential, commercial and utility-scale PV systems for eight major markets as of 2013 *(see table)*. However, DOE's SunShot Initiative has reported much lower U.S. installation prices. In 2014, prices continued to decline. The SunShot

Initiative modeled U.S. system prices to be in the range of $1.80 to $3.29 per watt. Other sources identify similar price ranges of $1.70 to $3.50 for the different market segments in the U.S., and in the highly penetrated German market, prices for residential and small commercial rooftop systems of up to 100 kW declined to $1.36 per watt (€1.24/W) by the end of 2014. In 2015, Deutsche Bank estimated costs for small residential rooftop systems in the U.S. around $2.90 per watt. Costs for utility-scale systems in China and India were estimated as low as $1.00 per watt.

Typical PV system prices in 2013 in selected countries (USD)								
USD/W	Australia	China	France	Germany	Italy	Japan	United Kingdom	United States
Residential	1.8	1.5	4.1	2.4	2.8	4.2	2.8	4.9[1]
Commercial	1.7	1.4	2.7	1.8	1.9	3.6	2.4	4.5[1]
Utility-scale	2.0	1.4	2.2	1.4	1.5	2.9	1.9	3.3[1]
Source: *IEA – Technology Roadmap: Solar Photovoltaic Energy report, September 2014'* [1]U.S figures are lower in DOE's Photovoltaic System Pricing Trends								

Grid Parity

Grid parity, the point at which the cost of photovoltaic electricity is equal to or cheaper than the price of grid power, is more easily achieved in areas with abundant sun and high costs for electricity such as in California and Japan. In 2008, The levelized cost of electricity for solar PV was $0.25/kWh or less in most of the OECD countries. By late 2011, the fully loaded cost was predicted to fall below $0.15/kWh for most of the OECD and to reach $0.10/kWh in sunnier regions. These cost levels are driving three emerging trends: vertical integration of the supply chain, origination of power purchase agreements (PPAs) by solar power companies, and unexpected risk for traditional power generation companies, grid operators and wind turbine manufacturers.

Grid parity was first reached in Spain in 2013, Hawaii and other islands that otherwise use fossil fuel (diesel fuel) to produce electricity, and most of the US is expected to reach grid parity by 2015. In 2007, General Electric's Chief Engineer predicted grid parity without subsidies in sunny parts of the United States by around 2015; other companies predicted an earlier date: the cost of solar power will be below grid parity for more than half of residential customers and 10% of commercial customers in the OECD, as long as grid electricity prices do not decrease through 2010.

Self Consumption

In cases of self consumption of the solar energy, the payback time is calculated based on how much electricity is not purchased from the grid. For example, in Germany, with electricity prices of 0.25 Euro/KWh and insolation of 900 KWh/KW, one KWp will save 225 Euro per year, and with an installation cost of 1700 Euro/KWp the system cost will be returned in less than 7 years. However, in many cases, the patterns of generation and consumption do not coincide, and some or all of the energy is fed back into the grid. The electricity is sold, and at other times when energy is taken from the grid, electricity is bought. The relative costs and prices obtained affect the economics.

Energy Pricing and Incentives

The political purpose of incentive policies for PV is to facilitate an initial small-scale deployment to begin to grow the industry, even where the cost of PV is significantly above grid parity, to allow the industry to achieve the economies of scale necessary to reach grid parity. The policies are implemented to promote national energy independence, high tech job creation and reduction of CO_2 emissions. Three incentive mechanisms are often used in combination as investment subsidies: the authorities refund part of the cost of installation of the system, the electricity utility buys PV electricity from the producer under a multiyear contract at a guaranteed rate (), and Solar Renewable Energy Certificates (SRECs)

Rebates

With investment subsidies, the financial burden falls upon the taxpayer, while with feed-in tariffs the extra cost is distributed across the utilities' customer bases. While the investment subsidy may be simpler to administer, the main argument in favour of feed-in tariffs is the encouragement of quality. Investment subsidies are paid out as a function of the nameplate capacity of the installed system and are independent of its actual power yield over time, thus rewarding the overstatement of power and tolerating poor durability and maintenance. Some electric companies offer rebates to their customers, such as Austin Energy in Texas, which offers $2.50/watt installed up to $15,000.

Net Metering

In net metering the price of the electricity produced is the same as the price supplied to the consumer, and the consumer is billed on the difference between production and consumption. Net metering can usually be done with no changes to standard electricity meters, which accurately measure power in both directions and automatically report the difference, and because it allows homeowners and businesses to generate electricity at a different time from consumption, effectively using the grid as a giant storage battery. With net metering, deficits are billed each month while surpluses are rolled over to the following month. Best practices call for perpetual roll over of kWh credits. Excess credits upon termination of service are either lost, or paid for at a rate ranging from wholesale to retail rate or above, as can be excess annual credits. In New Jersey, annual excess credits are paid at the wholesale rate, as are left over credits when a customer terminates service.

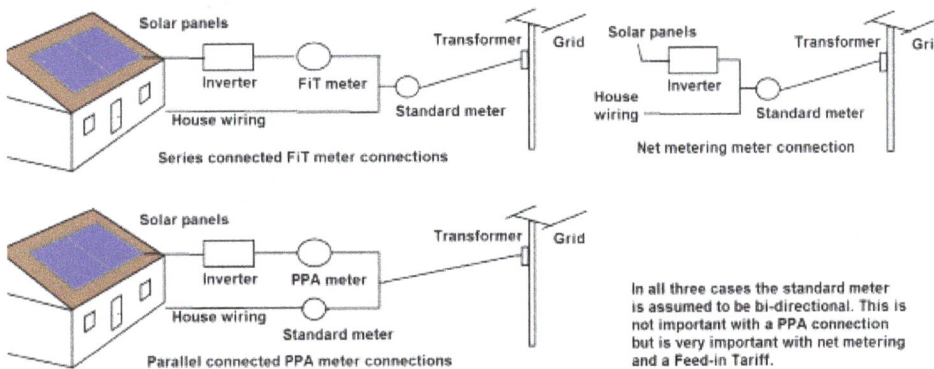

Understanding Feed-in Tariff and Power Purchase Agreement meter connections

Net metering, unlike a feed-in tariff, requires only one meter, but it must be bi-directional.

Feed-In Tariffs (FIT)

With feed-in tariffs, the financial burden falls upon the consumer. They reward the number of kilowatt-hours produced over a long period of time, but because the rate is set by the authorities, it may result in perceived overpayment. The price paid per kilowatt-hour under a feed-in tariff exceeds the price of grid electricity. Net metering refers to the case where the price paid by the utility is the same as the price charged.

The complexity of approvals in California, Spain and Italy has prevented comparable growth to Germany even though the return on investment is better. In some countries, additional incentives are offered for BIPV compared to stand alone PV.

- France + EUR 0.16 /kWh (compared to semi-integrated) or + EUR 0.27/kWh (compared to stand alone)

- Italy + EUR 0.04-0.09 kWh

- Germany + EUR 0.05/kWh (facades only)

Solar Renewable Energy Credits (SRECs)

Alternatively, SRECs allow for a market mechanism to set the price of the solar generated electricity subsidy. In this mechanism, a renewable energy production or consumption target is set, and the utility (more technically the Load Serving Entity) is obliged to purchase renewable energy or face a fine (Alternative Compliance Payment or ACP). The producer is credited for an SREC for every 1,000 kWh of electricity produced. If the utility buys this SREC and retires it, they avoid paying the ACP. In principle this system delivers the cheapest renewable energy, since the all solar facilities are eligible and can be installed in the most economic locations. Uncertainties about the future value of SRECs have led to long-term SREC contract markets to give clarity to their prices and allow solar developers to pre-sell and hedge their credits.

Financial incentives for photovoltaics differ across countries, including Australia, China, Germany, Israel, Japan, and the United States and even across states within the US.

The Japanese government through its Ministry of International Trade and Industry ran a successful programme of subsidies from 1994 to 2003. By the end of 2004, Japan led the world in installed PV capacity with over 1.1 GW.

In 2004, the German government introduced the first large-scale feed-in tariff system, under the German Renewable Energy Act, which resulted in explosive growth of PV installations in Germany. At the outset the FIT was over 3x the retail price or 8x the industrial price. The principle behind the German system is a 20-year flat rate contract. The value of new contracts is programmed to decrease each year, in order to encourage the industry to pass on lower costs to the end users. The programme has been more successful than expected with over 1GW installed in 2006, and political pressure is mounting to decrease the tariff to lessen the future burden on consumers.

Subsequently, Spain, Italy, Greece—that enjoyed an early success with domestic solar-thermal installations for hot water needs—and France introduced feed-in tariffs. None have replicated the programmed decrease of FIT in new contracts though, making the German incentive relatively less and less attractive compared to other countries. The French and Greek FIT offer a high premium (EUR 0.55/kWh) for building integrated systems. California, Greece, France and Italy have 30-50% more insolation than Germany making them financially more attractive. The Greek domestic "solar roof" programme (adopted in June 2009 for installations up to 10 kW) has internal rates of return of 10-15% at current commercial installation costs, which, furthermore, is tax free.

In 2006 California approved the 'California Solar Initiative', offering a choice of investment subsidies or FIT for small and medium systems and a FIT for large systems. The small-system FIT of $0.39 per kWh (far less than EU countries) expires in just 5 years, and the alternate "EPBB" residential investment incentive is modest, averaging perhaps 20% of cost. All California incentives are scheduled to decrease in the future depending as a function of the amount of PV capacity installed.

At the end of 2006, the Ontario Power Authority (OPA, Canada) began its Standard Offer Program, a precursor to the Green Energy Act, and the first in North America for distributed renewable projects of less than 10 MW. The feed-in tariff guaranteed a fixed price of $0.42 CDN per kWh over a period of twenty years. Unlike net metering, all the electricity produced was sold to the OPA at the given rate.

Environmental Impacts

Unlike fossil fuel based technologies, solar power does not lead to any harmful emissions during operation, but the production of the panels leads to some amount of pollution.

Part of the Senftenberg Solarpark, a solar photovoltaic power plant located on former open-pit mining areas close to the city of Senftenberg, in Eastern Germany. The 78 MW Phase 1 of the plant was completed within three months.

Greenhouse Gases

The Life-cycle greenhouse-gas emissions of solar power are in the range of 22 to 46 gram (g) per kilowatt-hour (kWh) depending on if solar thermal or solar PV is being analyzed, respectively. With this potentially being decreased to 15 g/kWh in the future. For comparison (of weighted averages), a combined cycle gas-fired power plant emits some 400–599 g/kWh, an oil-fired power plant 893 g/kWh, a coal-fired power plant 915–994 g/kWh or with carbon capture and storage some 200 g/kWh, and a geothermal high-temp. power plant 91–122 g/kWh. The life cycle emission intensity of hydro, wind and nuclear power are lower than solar's as of 2011 as published by the IPCC, and discussed in the article Life-cycle greenhouse-gas emissions of energy sources. Similar to all energy sources were their total life cycle emissions primarily lay in the construction and transportation phase, the switch to low carbon power in the manufacturing and transportation of solar devices would further reduce carbon emissions. BP Solar owns two factories built by Solarex (one in Maryland, the other in Virginia) in which all of the energy used to manufacture solar panels is produced by solar panels. A 1-kilowatt system eliminates the burning of approximately 170 pounds of coal, 300 pounds of carbon dioxide from being released into the atmosphere, and saves up to 105 gallons of water consumption monthly.

The US National Renewable Energy Laboratory (NREL), in harmonizing the disparate estimates of life-cycle GHG emissions for solar PV, found that the most critical parameter was the solar insolation of the site: GHG emissions factors for PV solar are inversely proportional to insolation. For a site with insolation of 1700 kWh/m2/year, typical of southern Europe, NREL researchers estimated GHG emissions of 45 gCO_2e/kWh. Using the same assumptions, at Phoenix, USA, with insolation of 2400 kWh/m2/year, the GHG emissions factor would be reduced to 32 g of CO_2e/kWh.

The New Zealand Parliamentary Commissioner for the Environment found that the solar PV would have little impact on the country's greenhouse gas emissions. The country already generates 80 percent of its electricity from renewable resources (primarily hydroelectricity and geothermal) and national electricity usage peaks on winter evenings whereas solar generation peaks on summer afternoons, meaning a large uptake of solar PV would end up displacing other renewable generators before fossil-fueled power plants.

Energy Payback

The energy payback time (EPBT) of a power generating system is the time required to generate as much energy as is consumed during production and lifetime operation of the system. Due to improving production technologies the payback time has been decreasing constantly since the introduction of PV systems in the energy market. In 2000 the energy payback time of PV systems was estimated as 8 to 11 years and in 2006 this was estimated to be 1.5 to 3.5 years for crystalline silicon silicon PV systems and 1–1.5 years for thin film technologies (S. Europe). These figures fell to 0.75–3.5 years in 2013, with an average of about 2 years for crystalline silicon PV and CIS systems.

Another economic measure, closely related to the energy payback time, is the energy returned on energy invested (EROEI) or energy return on investment (EROI), which is the ratio of electricity generated divided by the energy required to build *and maintain* the equipment. (This is not the same as the economic return on investment (ROI), which varies according to local energy prices, subsidies available and metering techniques.) With expected lifetimes of 30 years, the EROEI of PV systems are in the range of 10 to 30, thus generating enough energy over their lifetimes to reproduce themselves many times (6-31 reproductions) depending on what type of material, balance of system (BOS), and the geographic location of the system.

Other Issues

One issue that has often raised concerns is the use of cadmium (Cd), a toxic heavy metal that has the tendency to accumulate in ecological food chains. It is used as semiconductor component in

CdTe solar cells and as buffer layer for certain CIGS cells in the form of CdS. The amount of cadmium used in thin-film PV modules is relatively small (5–10 g/m^2) and with proper recycling and emission control techniques in place the cadmium emissions from module production can be almost zero. Current PV technologies lead to cadmium emissions of 0.3–0.9 microgram/kWh over the whole life-cycle. Most of these emissions actually arise through the use of coal power for the manufacturing of the modules, and coal and lignite combustion leads to much higher emissions of cadmium. Life-cycle cadmium emissions from coal is 3.1 microgram/kWh, lignite 6.2, and natural gas 0.2 microgram/kWh.

In a life-cycle analysis it has been noted, that if electricity produced by photovoltaic panels were used to manufacture the modules instead of electricity from burning coal, cadmium emissions from coal power usage in the manufacturing process could be entirely eliminated.

In the case of crystalline silicon modules, the solder material, that joins together the copper strings of the cells, contains about 36 percent of lead (Pb). Moreover, the paste used for screen printing front and back contacts contains traces of Pb and sometimes Cd as well. It is estimated that about 1,000 metric tonnes of Pb have been used for 100 gigawatts of c-Si solar modules. However, there is no fundamental need for lead in the solder alloy.

Some media sources have reported that concentrated solar power plants have injured or killed large numbers of birds due to intense heat from the concentrated sunrays. This adverse effect does not apply to PV solar power plants, and some of the claims may have been overstated or exaggerated.

A 2014-published life-cycle analysis of land use for various sources of electricity concluded that the large-scale implementation of solar and wind potentially reduces pollution-related environmental impacts. The study found that the land-use footprint, given in square meter-years per megawatt-hour (m^{2}a/MWh), was lowest for wind, natural gas and rooftop PV, with 0.26, 0.49 and 0.59, respectively, and followed by utility-scale solar PV with 7.9. For CSP, the footprint was 9 and 14, using parabolic troughs and solar towers, respectively. The largest footprint had coal-fired power plants with 18 m^{2}a/MWh.

Emerging Technologies

Concentrator Photovoltaics

CPV modules on dual axis solar trackers in Golmud, China

Concentrator photovoltaics (CPV) systems employ sunlight concentrated onto photovoltaic surfaces for the purpose of electrical power production. Contrary to conventional photovoltaic systems, it uses lenses and curved mirrors to focus sunlight onto small, but highly efficient, multi-junction solar cells. Solar concentrators of all varieties may be used, and these are often mounted on a solar tracker in order to keep the focal point upon the cell as the sun moves across the sky. Luminescent solar concentrators (when combined with a PV-solar cell) can also be regarded as a CPV system. Concentrated photovoltaics are useful as they can improve efficiency of PV-solar panels drastically.

In addition, most solar panels on spacecraft are also made of high efficient multi-junction photovoltaic cells to derive electricity from sunlight when operating in the inner Solar System.

Floatovoltaics

Floatovoltaics are an emerging form of PV systems that float on the surface of irrigation canals, water reservoirs, quarry lakes, and tailing ponds. Several systems exist in France, India, Japan, Korea, the United Kingdom and the United States. These systems reduce the need of valuable land area, save drinking water that would otherwise be lost through evaporation, and show a higher efficiency of solar energy conversion, as the panels are kept at a cooler temperature than they would be on land.

Grid Integration

Since solar energy is not available at night, storing its energy is an important issue in order to have continuous energy availability. Both wind power and solar power are variable renewable energy, meaning that all available output must be taken when it is available, and either stored for *when it can be used later*, or transported over transmission lines to *where it can be used now*. Concentrated solar power plants typically use thermal energy storage to store the solar energy, such as in high-temperature molten salts. These salts are an effective storage medium because they are low-cost, have a high specific heat capacity, and can deliver heat at temperatures compatible with conventional power systems. This method of energy storage is used, for example, by the Solar Two power station, allowing it to store 1.44 TJ in its 68 m³ storage tank, enough to provide full output for close to 39 hours, with an efficiency of about 99%.

Construction of the Salt Tanks which provide efficient thermal energy storage so that output can be provided after the sun goes down, and output can be scheduled to meet demand requirements. The 280 MW Solana Generating Station is designed to provide six hours of energy storage. This allows the plant to generate about 38 percent of its rated capacity over the course of a year.

Thermal energy storage. The Andasol CSP plant uses tanks of molten salt to store solar energy.

Rechargeable batteries have been traditionally used to store excess electricity in stand alone PV systems. With grid-connected photovoltaic power system, excess electricity can be sent to the electrical grid. Net metering and feed-in tariff programs give these systems a credit for the electricity they produce. This credit offsets electricity provided from the grid when the system cannot meet demand, effectively using the grid as a storage mechanism. Credits are normally rolled over from month to month and any remaining surplus settled annually. When wind and solar are a small fraction of the grid power, other generation techniques can adjust their output appropriately, but as these forms of variable power grow, this becomes less practical. As prices are rapidly declining, PV systems increasingly use rechargeable batteries to store a surplus to be later used at night. Batteries used for grid-storage also stabilize the electrical grid by leveling out peak loads, and play an important role in a smart grid, as they can charge during periods of low demand and feed their stored energy into the grid when demand is high.

Pumped-storage hydroelectricity (PSH). This facility in Geesthacht, Germany, also includes a solar array.

Like plug in cars, it is technically possible to have "plug and play" PV. A recent review article found that careful system design would enable such systems to meet all technical and safety requirements.

Common battery technologies used in today's PV systems include, the valve regulated lead-acid battery– a modified version of the conventional lead–acid battery, nickel–cadmium and lithium-ion batteries. Lead-acid batteries are currently the predominant technology used in small-scale, residential PV systems, due to their high reliability, low self discharge and investment and maintenance costs, despite shorter lifetime and lower energy density. However, lithium-ion batteries have the potential to replace lead-acid batteries in the near future, as they are being intensively developed and lower prices are expected due to economies of scale provided by large production facilities such as the Gigafactory 1. In addition, the Li-ion batteries of plug-in electric cars may serve as a future storage devices in a vehicle-to-grid system. Since most vehicles are parked an average of 95 percent of the time, their batteries could be used to let electricity flow from the car to the power lines and back. Other rechargeable batteries used for distributed PV systems include, sodium–sulfur and vanadium redox batteries, two prominent types of a molten salt and a flow battery, respectively.

Conventional hydroelectricity works very well in conjunction with variable electricity sources such as solar and wind, the water can be held back and allowed to flow as required with virtually no energy loss. Where a suitable river is not available, pumped-storage hydroelectricity stores energy in the form of water pumped when surplus electricity is available, from a lower elevation reservoir to a higher elevation one. The energy is recovered when demand is high by releasing the water: the pump becomes a turbine, and the motor a hydroelectric power generator. However, this loses some of the energy to pumpage losses.

The combination of wind and solar PV has the advantage that the two sources complement each other because the peak operating times for each system occur at different times of the day and year. The power generation of such solar hybrid power systems is therefore more constant and fluctuates less than each of the two component subsystems. Solar power is seasonal, particularly in northern/southern climates, away from the equator, suggesting a need for long term seasonal storage in a medium such as hydrogen. The storage requirements vary and in some cases can be met with biomass. The Institute for Solar Energy Supply Technology of the University of Kassel pilot-tested a combined power plant linking solar, wind, biogas and hydrostorage to provide load-following power around the clock, entirely from renewable sources.

Research is also undertaken in this field of artificial photosynthesis. It involves the use of nanotechnology to store solar electromagnetic energy in chemical bonds, by splitting water to produce hydrogen fuel or then combining with carbon dioxide to make biopolymers such as methanol. Many large national and regional research projects on artificial photosynthesis are now trying to develop techniques integrating improved light capture, quantum coherence methods of electron transfer and cheap catalytic materials that operate under a variety of atmospheric conditions. Senior researchers in the field have made the public policy case for a Global Project on Artificial Photosynthesis to address critical energy security and environmental sustainability issues.

Geographic Solar Insolation

Different parts of the world experience different amounts of sunshine, depending on latitude and weather. Locations nearer the equator receive many more hours of sunshine than those further north or south, thus photovoltaic panels can be more economically desirable in some places more than others.

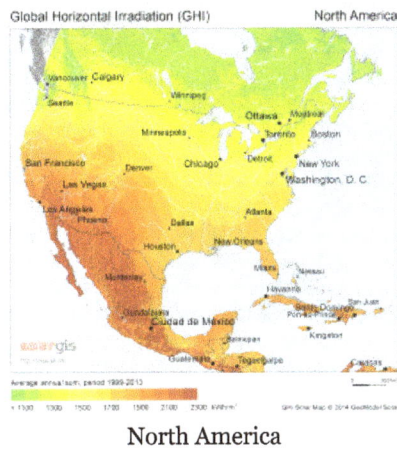

North America

South America

Photovoltaics

Photovoltaics (PV) covers the conversion of light into electricity using semiconducting materials that exhibit the photovoltaic effect, a phenomenon studied in physics, photochemistry, and electrochemistry.

The Solar Settlement, a sustainable housing community project in Freiburg, Germany.

Double glass photovoltaic solar modules, installed in a support structure.

A typical photovoltaic system employs solar panels, each comprising a number of solar cells, which generate electrical power. The first step is the photoelectric effect followed by an electrochemical process where crystallized atoms, ionized in a series, generate an electric current. PV Installations may be ground-mounted, rooftop mounted or wall mounted.

Photovoltaic SUDI shade is an autonomous and mobile station in France that provides energy for electric vehicles using solar energy.

Solar PV generates no pollution. The direct conversion of sunlight to electricity occurs without any moving parts. Photovoltaic systems have been used for fifty years in specialized applications, standalone and grid-connected PV systems have been in use for more than twenty years. They were first mass-produced in 2000, when German environmentalists and the Eurosolar organization got government funding for a ten thousand roof program.

Solar panels on the International Space Station

On the other hand, grid-connected PV systems have the major disadvantage that the power output is dependent on direct sunlight, so about 10-25% is lost if a tracking system is not used, since the cell wil not be directly facing the sun at all times.Power output is also adversely affected by weather conditions, especially cloud cover. This means that, in the national grid for example, this power has to be made up by other power sources: hydrocarbon, nuclear, hydroelectric or wind energy. To some, solar installations also have a negative aesthetic impact on an area.

Advances in technology and increased manufacturing scale have reduced the cost, increased the reliability, and increased the efficiency of photovoltaic instalations and the levelised cost of electricity from PV is competitive, on a kilowatt/ hour basis, with conventional electricity sources in an expanding list of geographic regions. Solar PV regularly costs USD 0.05-0.10 per kilowatt-hour (kWh) in Europe, China, India, South Africa and the United States. In 2015, record low prices were set in the United Arab Emirates (5.84 cents/kWh), Peru (4.8 cents/kWh) and Mexico (4.8 cents/kWh). In May 2016, a solar PV auction in Dubai attracted a bid of 3 cents/kWh.

Net metering and financial incentives, such as preferential feed-in tariffs for solar-generated electricity, have supported solar PV installations in many countries. More than 100 countries now use solar PV. After hydro and wind power, PV is the third renewable energy source in terms of globally capacity. In 2014, worldwide installed PV capacity increased to 177 gigawatts (GW), which is two percent of global electricity demand. China, followed by Japan and the United States, is the fastest growing market, while Germany remains the world's largest producer (both in per capita and absolute terms), with solar PV providing seven percent of annual domestic electricity consumption.

With current technology (as of 2013), photovoltaics recoups the energy needed to manufacture them in 1.5 years in Southern Europe and 2.5 years in Northern Europe.

Etymology

The term "photovoltaic" meaning "light", and from "volt", the unit of electro-motive force, the volt, which in turn comes from the last name of the Italian physicist Alessandro Volta, inventor of the battery (electrochemical cell). The term "photo-voltaic" has been in use in English since 1849.

Solar Cells

Solar cells generate electricity directly from sunlight.

Photovoltaics are best known as a method for generating electric power by using solar cells to convert energy from the sun into a flow of electrons. The photovoltaic effect refers to photons of light exciting electrons into a higher state of energy, allowing them to act as charge carriers for an electric current. The photovoltaic effect was first observed by Alexandre-Edmond Becquerel in 1839. The term photovoltaic denotes the unbiased operating mode of a photodiode in which current through the device is entirely due to the transduced light energy. Virtually all photovoltaic devices are some type of photodiode.

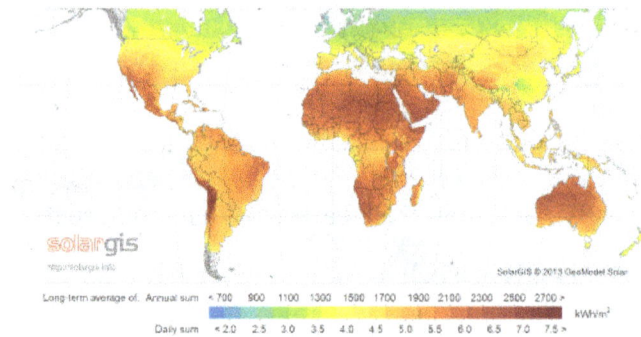

Average insolation. Note that this is for a horizontal surface. Solar panels are normally propped up at an angle and receive more energy per unit area.

Solar cells produce direct current electricity from sun light which can be used to power equipment or to recharge a battery. The first practical application of photovoltaics was to power orbiting satellites and other spacecraft, but today the majority of photovoltaic modules are used for grid connected power generation. In this case an inverter is required to convert the DC to AC. There is a smaller market for off-grid power for remote dwellings, boats, recreational vehicles, electric cars, roadside emergency telephones, remote sensing, and cathodic protection of pipelines.

Photovoltaic power generation employs solar panels composed of a number of solar cells containing a photovoltaic material. Materials presently used for photovoltaics include monocrystalline silicon, polycrystalline silicon, amorphous silicon, cadmium telluride, and copper indium gallium selenide/sulfide. Copper solar cables connect modules (module cable), arrays (array cable), and sub-fields. Because of the growing demand for renewable energy sources, the manufacturing of solar cells and photovoltaic arrays has advanced considerably in recent years.

Solar photovoltaics power generation has long been seen as a clean energy technology which draws upon the planet's most plentiful and widely distributed renewable energy source – the sun. The technology is "inherently elegant" in that the direct conversion of sunlight to electricity occurs without any moving parts or environmental emissions during operation. It is well proven, as photovoltaic systems have now been used for fifty years in specialised applications, and grid-connected systems have been in use for over twenty years.

Cells require protection from the environment and are usually packaged tightly behind a glass sheet. When more power is required than a single cell can deliver, cells are electrically connected together to form photovoltaic modules, or solar panels. A single module is enough to power an emergency telephone, but for a house or a power plant the modules must be arranged in multiples as arrays.

Photovoltaic power capacity is measured as maximum power output under standardized test conditions (STC) in "W_p" (watts peak). The actual power output at a particular point in time may be less than or greater than this standardized, or "rated," value, depending on geographical location, time of day, weather conditions, and other factors. Solar photovoltaic array capacity factors are typically under 25%, which is lower than many other industrial sources of electricity.

Current Developments

For best performance, terrestrial PV systems aim to maximize the time they face the sun. Solar trackers achieve this by moving PV panels to follow the sun. The increase can be by as much as 20% in winter and by as much as 50% in summer. Static mounted systems can be optimized by analysis of the sun path. Panels are often set to latitude tilt, an angle equal to the latitude, but performance can be improved by adjusting the angle for summer or winter. Generally, as with other semiconductor devices, temperatures above room temperature reduce the performance of photovoltaics.

A number of solar panels may also be mounted vertically above each other in a tower, if the zenith distance of the Sun is greater than zero, and the tower can be turned horizontally as a whole and each panels additionally around a horizontal axis. In such a tower the panels can follow the Sun exactly. Such a device may be described as a ladder mounted on a turnable disk. Each step of that ladder is the middle axis of a rectangular solar panel. In case the zenith distance of the Sun reaches zero, the "ladder" may be rotated to the north or the south to avoid a solar panel producing a shadow on a lower solar panel. Instead of an exactly vertical tower one can choose a tower with an axis directed to the polar star, meaning that it is parallel to the rotation axis of the Earth. In this case the angle between the axis and the Sun is always larger than 66 degrees. During a day it is only necessary to turn the panels around this axis to follow the Sun. Installations may be ground-mounted (and sometimes integrated with farming and grazing) or built into the roof or walls of a building (building-integrated photovoltaics).

Another recent development involves the makeup of solar cells. Perovskite is a very inexpensive material which is being used to replace the expensive crystalline silicon which is still part of a standard PV cell build to this day. Michael Graetzel, Director of the Laboratory of Photonics and Interfaces at EPFL says, "Today, efficiency has peaked at 18 percent, but it's expected to get even higher in the future." This is a significant claim, as 20% efficiency is typical among solar panels which use more expensive materials.

Efficiency

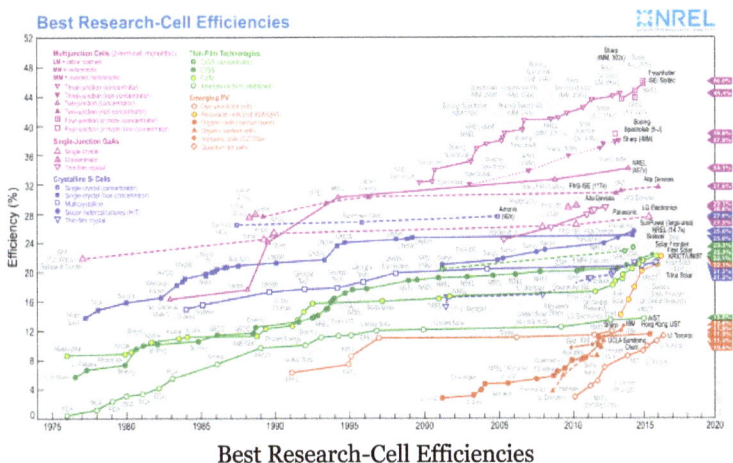

Best Research-Cell Efficiencies

Electrical efficiency (also called conversion efficiency) is a contributing factor in the selection of a photovoltaic system. However, the most efficient solar panels are typically the most expensive, and may not be commercially available. Therefore, selection is also driven by cost efficiency and other factors.

The electrical efficiency of a PV cell is a physical property which represents how much electrical power a cell can produce for a given insolation. The basic expression for maximum efficiency of a photovoltaic cell is given by the ratio of output power to the incident solar power (radiation flux times area)

$$\eta = \frac{P_{max}}{E \cdot A_{cell}}.$$

The efficiency is measured under ideal laboratory conditions and represents the maximum achievable efficiency of the PV material. Actual efficiency is influenced by the output Voltage, current, junction temperature, light intensity and spectrum.

The most efficient type of solar cell to date is a multi-junction concentrator solar cell with an efficiency of 46.0% produced by Fraunhofer ISE in December 2014. The highest efficiencies achieved without concentration include a material by Sharp Corporation at 35.8% using a proprietary triple-junction manufacturing technology in 2009, and Boeing Spectrolab (40.7% also using a triple-layer design). The US company SunPower produces cells that have an efficiency of 21.5%, well above the market average of 12–18%.

There is an ongoing effort to increase the conversion efficiency of PV cells and modules, primarily for competitive advantage. In order to increase the efficiency of solar cells, it is important to choose a semiconductor material with an appropriate band gap that matches the solar spectrum. This will enhance the electrical and optical properties. Improving the method of charge collection is also useful for increasing the efficiency. There are several groups of materials that are being developed. Ultrahigh-efficiency devices ($\eta > 30\%$) are made by using GaAs and GaInP2 semiconductors with multijunction tandem cells. High-quality, single-crystal silicon materials are used to achieve high-efficiency, low cost cells ($\eta > 20\%$).

Recent developments in Organic photovoltaic cells (OPVs) have made significant advancements in power conversion efficiency from 3% to over 15% since their introduction in the 1980s. To date, the highest reported power conversion efficiency ranges from 6.7% to 8.94% for small molecule, 8.4%–10.6% for polymer OPVs, and 7% to 21% for perovskite OPVs. OPV's are expected to play a major role in the PV market. Recent improvements have increased the efficiency and lowered cost, while remaining environmentally-benign and renewable.

Several companies have begun embedding power optimizers into PV modules called smart modules. These modules perform maximum power point tracking (MPPT) for each module individually, measure performance data for monitoring, and provide additional safety features. Such modules can also compensate for shading effects, wherein a shadow falling across a section of a module causes the electrical output of one or more strings of cells in the module to decrease.

One of the major causes for the decreased performance of cells is overheating. The efficiency of a solar cell declines by about 0.5% for every 1 degree Celsius increase in temperature. This means that a

100 degree increase in surface temperature could decrease the efficiency of a solar cell by about half. Self-cooling solar cells are one solution to this problem. Rather than using energy to cool the surface, pyramid and cone shapes can be formed from silica, and attached to the surface of a solar panel. Doing so allows visible light to reach the solar cells, but reflects infrared rays (which carry heat).

Growth

Projected Global Growth (MW)

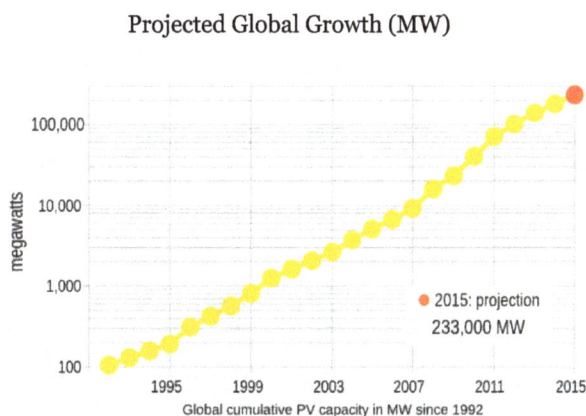

Worldwide growth of photovoltaics on a semi-log plot since 1992

Solar photovoltaics is growing rapidly and worldwide installed capacity reached at least 177 gigawatts (GW) by the end of 2014. The total power output of the world's PV capacity in a calendar year is now beyond 200 TWh of electricity. This represents 1% of worldwide electricity demand. More than 100 countries use solar PV. China, followed by Japan and the United States is now the fastest growing market, while Germany remains the world's largest producer, contributing more than 7% to its national electricity demands. Photovoltaics is now, after hydro and wind power, the third most important renewable energy source in terms of globally installed capacity.

Several market research and financial companies foresee record-breaking global installation of more than 50 GW in 2015. China is predicted to take the lead from Germany and to become the world's largest producer of PV power by installing another targeted 17.8 GW in 2015. India is expected to install 1.8 GW, doubling its annual installations. By 2018, worldwide photovoltaic capacity is projected to doubled or even triple to 430 GW. Solar Power Europe (formerly known as EPIA) also estimates that photovoltaics will meet 10% to 15% of Europe's energy demand in 2030.

The EPIA/Greenpeace Solar Generation Paradigm Shift Scenario (formerly called Advanced Scenario) from 2010 shows that by the year 2030, 1,845 GW of PV systems could be generating approximately 2,646 TWh/year of electricity around the world. Combined with energy use efficiency improvements, this would represent the electricity needs of more than 9% of the world's population. By 2050, over 20% of all electricity could be provided by photovoltaics.

Michael Liebreich, from Bloomberg New Energy Finance, anticipates a tipping point for solar energy. The costs of power from wind and solar are already below those of conventional electricity generation in some parts of the world, as they have fallen sharply and will continue to do so. He also asserts, that the electrical grid has been greatly expanded worldwide, and is ready to receive and distribute electricity from renewable sources. In addition, worldwide electricity prices came under strong pressure from renewable energy sources, that are, in part, enthusiastically embraced by consumers.

Deutsche Bank sees a "second gold rush" for the photovoltaic industry to come. Grid parity has already been reached in at least 19 markets by January 2014. Photovoltaics will prevail beyond feed-in tariffs, becoming more competitive as deployment increases and prices continue to fall.

In June 2014 Barclays downgraded bonds of U.S. utility companies. Barclays expects more competition by a growing self-consumption due to a combination of decentralized PV-systems and residential electricity storage. This could fundamentally change the utility's business model and transform the system over the next ten years, as prices for these systems are predicted to fall.

Top 10 PV countries in 2014 (MW)					
Total capacity			**Added capacity**		
1.	Germany	38,200	1.	China	10,560
2.	China	28,199	2.	Japan	9,700
3.	Japan	23,300	3.	United States	6,201
4.	Italy	18,460	4.	UK	2,273
5.	United States	18,280	5.	Germany	1,900
6.	France	5,660	6.	France	927
7.	Spain	5,358	7.	Australia	910
8.	UK	5,104	8.	South Korea	909
9.	Australia	4,136	9.	South Africa	800
10.	Belgium	3,074	10.	India	616

Data: IEA-PVPS *Snapshot of Global PV 1992–2014* report, March 2015

Environmental Impacts of Photovoltaic Technologies

Types of Impacts

While solar photovoltaic (PV) cells are promising for clean energy production, their deployment is hindered by production costs, material availability, and toxicity. Life cycle assessment (LCA) is one method of determining environmental impacts from PV. Many studies have been done on the various types of PV including first generation, second generation, and third generation. Usually these PV LCA studies select a cradle to gate system boundary because often at the time the studies are conducted, it is a new technology not commercially available yet and their required balance of system components and disposal methods are unknown.

A traditional LCA can look at many different impact categories ranging from global warming potential, eco-toxicity, human toxicity, water depletion, and many others. Most LCAs of PV have focused on two categories: carbon dioxide equivalents per kWh and energy pay-back time (EPBT). The EPBT is defined as " the time needed to compensate for the total renewable- and non-renewable- primary energy required during the life cycle of a PV system". A 2015 review of EPBT from first and second generation PV suggested that there was greater variation in embedded energy than in efficiency of the cells implying that it was mainly the embedded energy that needs to reduce to have a greater reduction in EPBT. One difficulty in determining impacts due to PV is to determine if the wastes are released to the air, water, or soil during the manufacturing phase. Research is underway to try to understand emissions and releases during the lifetime of PV systems.

Impacts from First-generation PV

Crystalline silicon modules are the most extensively studied PV type in terms of LCA since they are the most commonly used. Mono-crystalline silicon photovoltaic systems (mono-si) have an average efficiency of 14.0%. The cells tend to follow a structure of front electrode, anti-reflection film, n-layer, p-layer, and back electrode, with the sun hitting the front electrode. EPBT ranges from 1.7 to 2.7 years. The cradle to gate of CO_2-eq/kWh ranges from 37.3 to 72.2 grams.

Techniques to produce multi-crystalline silicon (multi-si) photovoltaic cells are simpler and cheaper than mono-si, however tend to make less efficient cells, an average of 13.2%. EPBT ranges from 1.5 to 2.6 years. The cradle to gate of CO_2-eq/kWh ranges from 28.5 to 69 grams. Some studies have looked beyond EPBT and GWP to other environmental impacts. In one such study, conventional energy mix in Greece was compared to multi-si PV and found a 95% overall reduction in impacts including carcinogens, eco-toxicity, acidification, eutrophication, and eleven others.

Impacts from Second Generation

Cadmium telluride (CdTe) is one of the fastest-growing thin film based solar cells which are collectively known as second generation devices. This new thin film device also shares similar performance restrictions (Shockley-Queisser efficiency limit) as conventional Si devices but promises to lower the cost of each device by both reducing material and energy consumption during manufacturing. Today the global market share of CdTe is 5.4%, up from 4.7% in 2008. This technology's highest power conversion efficiency is 21%. The cell structure includes glass substrate (around 2 mm), transparent conductor layer, CdS buffer layer (50–150 nm), CdTe absorber and a metal contact layer.

CdTe PV systems require less energy input in their production than other commercial PV systems per unit electricity production. The average CO_2-eq/kWh is around 18 grams (cradle to gate). CdTe has the fastest EPBT of all commercial PV technologies, which varies between 0.3 and 1.2 years.

Copper Indium Gallium Diselenide (CIGS) is a thin film solar cell based on the copper indium diselenide (CIS) family of chalcopyrite semiconductors. CIS and CIGS are often used interchangeably within the CIS/CIGS community. The cell structure includes soda lime glass as the substrate, Mo layer as the back contact, CIS/CIGS as the absorber layer, cadmium sulfide (CdS) or Zn (S,OH) x as the buffer layer, and ZnO:Al as the front contact. CIGS is approximately 1/100th the thickness of conventional silicon solar cell technologies. Materials necessary for assembly are readily available, and are less costly per watt of solar cell. CIGS based solar devices resist performance degradation over time and are highly stable in the field.

Reported global warming potential impacts of CIGS range from 20.5 – 58.8 grams CO_2-eq/kWh of electricity generated for different solar irradiation (1,700 to 2,200 kWh/m²/y) and power conversion efficiency (7.8 – 9.12%). EPBT ranges from 0.2 to 1.4 years, while harmonized value of EPBT was found 1.393 years. Toxicity is an issue within the buffer layer of CIGS modules because it contains cadmium and gallium. CIS modules do not contain any heavy metals.

Impacts from Third Generation

Third-generation PVs are designed to combine the advantages of both the first and second generation devices and they do not have Shockley-Queisser efficiency limit, a theoretical limit for first and second generation PV cells. The thickness of a third generation device is less than 1 μm.

One emerging alternative and promising technology is based on an organic-inorganic hybrid solar cell made of methylammonium lead halide perovskites. Perovskite PV cells have progressed rapidly over the past few years and have become one of the most attractive areas for PV research. The cell structure includes a metal back contact (which can be made of Al, Au or Ag), a hole transfer layer (spiro-MeOTAD, P3HT, PTAA, CuSCN, CuI, or NiO), and absorber layer ($CH_3NH_3PbIxBr_3$-x, $CH_3NH_3PbIxCl_3$-x or $CH_3NH_3PbI_3$), an electron transport layer (TiO, ZnO, Al_2O_3 or SnO_2) and a top contact layer (fluorine doped tin oxide or tin doped indium oxide).

There are a limited number of published studies to address the environmental impacts of perovskite solar cells. The major environmental concern is the lead used in the absorber layer. Due to the instability of perovskite cells lead may eventually be exposed to fresh water during the use phase. Two published LCA studies looked at human and ecotoxicity of perovskite solar cells and found they were surprisingly low and may not be an environmental issue. Gong et al. found direct processing energy as 30 MJ/m², while Espinosa didn't report this value (but estimated around 1000 MJ/m²). Global warming potential was found to be in the range of 24–1500 grams CO_2-eq/kWh electricity production. Similarly, reported EPBT of the published paper range from 0.2 to 15 years. The large range of reported values highlight the uncertainties associated with these studies.

Two new promising thin film technologies are copper zinc tin sulfide (Cu_2ZnSnS_4 or CZTS) and zinc phosphide (Zn_3P_2). Both of these thin films are currently only produced in the lab but may be commercialized in the future. Their manufacturing processes are expected to be similar to those of current thin film technologies of CIGS and CdTe, respectively. Yet, contrary to CIGS and CdTe, CZTS and Zn_3P_2 are made from earth abundant, nontoxic materials and have the potential to produce more electricity annually than the current worldwide consumption. While CZTS and Zn_3P_2 offer good promise for these reasons, the specific environmental implications of their commercial production are not yet known. Global warming potential of CZTS and Zn_3P_2 were found 38 and 30 grams CO_2-eq/kWh while their corresponding EPBT were found 1.85 and 0.78 years, respectively. Overall, CdTe and Zn_3P_2 have similar environmental impacts but can slightly outperform CIGS and CZTS.

Organic and polymer photovoltaic (OPV) are a relatively new area of research. The tradition OPV cell structure layers consist of a semi-transparent electrode, electron blocking layer, tunnel junction, holes blocking layer, electrode, with the sun hitting the transparent electrode. OPV replaces silver with carbon as an electrode material lowering manufacturing cost and making them more environmentally friendly. OPV are flexible, low weight, and work well with roll-to roll manufacturing for mass production. OPV uses "only abundant elements coupled to an extremely low embodied energy through very low processing temperatures using only ambient processing conditions on simple printing equipment enabling energy pay-back times". Current efficiencies range from 1–6.5%, however theoretical analyses show promise beyond 10% efficiency.

Many different configurations of OPV exist using different materials for each layer. OPV technology rivals existing PV technologies in terms of EPBT even if they currently present a shorter operational lifetime. A 2013 study analyzed 12 different configurations all with 2% efficiency, the EPBT ranged from 0.29–0.52 years for 1 m² of PV. The average CO_2-eq/kWh for OPV is 54.922 grams.

Economics

There have been major changes in the underlying costs, industry structure and market prices of solar photovoltaics technology, over the years, and gaining a coherent picture of the shifts occurring across the industry value chain globally is a challenge. This is due to: "the rapidity of cost and price changes, the complexity of the PV supply chain, which involves a large number of manufacturing processes, the balance of system (BOS) and installation costs associated with complete PV systems, the choice of different distribution channels, and differences between regional markets within which PV is being deployed". Further complexities result from the many different policy support initiatives that have been put in place to facilitate photovoltaics commercialisation in various countries.

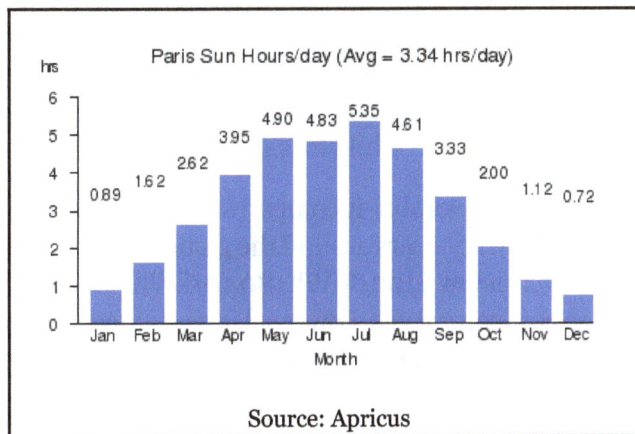

Source: Apricus

The PV industry has seen dramatic drops in module prices since 2008. In late 2011, factory-gate prices for crystalline-silicon photovoltaic modules dropped below the $1.00/W mark. The $1.00/W installed cost, is often regarded in the PV industry as marking the achievement of grid parity for PV. Technological advancements, manufacturing process improvements, and industry re-structuring, mean that further price reductions are likely in coming years.

Financial incentives for photovoltaics, such as feed-in tariffs, have often been offered to electricity consumers to install and operate solar-electric generating systems. Government has sometimes also offered incentives in order to encourage the PV industry to achieve the economies of scale needed to compete where the cost of PV-generated electricity is above the cost from the existing grid. Such policies are implemented to promote national or territorial energy independence, high tech job creation and reduction of carbon dioxide emissions which cause global warming. Due to economies of scale solar panels get less costly as people use and buy more—as manufacturers increase production to meet demand, the cost and price is expected to drop in the years to come.

Solar cell efficiencies vary from 6% for amorphous silicon-based solar cells to 44.0% with multiple-junction concentrated photovoltaics. Solar cell energy conversion efficiencies for commercially available photovoltaics are around 14–22%. Concentrated photovoltaics (CPV) may reduce cost by concentrating up to 1,000 suns (through magnifying lens) onto a smaller sized photovoltaic cell. However, such concentrated solar power requires sophisticated heat sink designs, otherwise the photovoltaic cell overheats, which reduces its efficiency and life. To further exacerbate the concentrated cooling design, the heat sink must be passive, otherwise the power required for active cooling would reduce the overall efficiency and economy.

Crystalline silicon solar cell prices have fallen from $76.67/Watt in 1977 to an estimated $0.74/Watt in 2013. This is seen as evidence supporting Swanson's law, an observation similar to the famous Moore's Law that states that solar cell prices fall 20% for every doubling of industry capacity.

As of 2011, the price of PV modules has fallen by 60% since the summer of 2008, according to Bloomberg New Energy Finance estimates, putting solar power for the first time on a competitive footing with the retail price of electricity in a number of sunny countries; an alternative and consistent price decline figure of 75% from 2007 to 2012 has also been published, though it is unclear whether these figures are specific to the United States or generally global. The levelised cost of electricity (LCOE) from PV is competitive with conventional electricity sources in an expanding list of geographic regions, particularly when the time of generation is included, as electricity is worth more during the day than at night. There has been fierce competition in the supply chain, and further improvements in the levelised cost of energy for solar lie ahead, posing a growing threat to the dominance of fossil fuel generation sources in the next few years. As time progresses, renewable energy technologies generally get cheaper, while fossil fuels generally get more expensive:

The less solar power costs, the more favorably it compares to conventional power, and the more attractive it becomes to utilities and energy users around the globe. Utility-scale solar power can now be delivered in California at prices well below $100/MWh ($0.10/kWh) less than most other peak generators, even those running on low-cost natural gas. Lower solar module costs also stimulate demand from consumer markets where the cost of solar compares very favorably to retail electric rates.

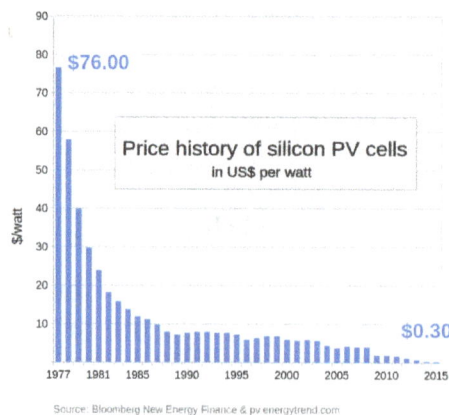

Price per watt history for conventional (c-Si) solar cells since 1977.

As of 2011, the cost of PV has fallen well below that of nuclear power and is set to fall further. The average retail price of solar cells as monitored by the Solarbuzz group fell from $3.50/watt to $2.43/watt over the course of 2011.

For large-scale installations, prices below $1.00/watt were achieved. A module price of 0.60 Euro/watt ($0.78/watt) was published for a large scale 5-year deal in April 2012.

By the end of 2012, the "best in class" module price had dropped to $0.50/watt, and was expected to drop to $0.36/watt by 2017.

In many locations, PV has reached grid parity, which is usually defined as PV production costs at or below retail electricity prices (though often still above the power station prices for coal or gas-fired generation without their distribution and other costs). However, in many countries there is still a need for more access to capital to develop PV projects. To solve this problem securitization has been proposed and used to accelerate development of solar photovoltaic projects. For example, SolarCity offered, the first U.S. asset-backed security in the solar industry in 2013.

Photovoltaic power is also generated during a time of day that is close to peak demand (precedes it) in electricity systems with high use of air conditioning. More generally, it is now evident that, given a carbon price of $50/ton, which would raise the price of coal-fired power by 5c/kWh, solar PV will be cost-competitive in most locations. The declining price of PV has been reflected in rapidly growing installations, totaling about 23 GW in 2011. Although some consolidation is likely in 2012, due to support cuts in the large markets of Germany and Italy, strong growth seems likely to continue for the rest of the decade. Already, by one estimate, total investment in renewables for 2011 exceeded investment in carbon-based electricity generation.

In the case of self consumption payback time is calculated based on how much electricity is not brought from the grid. Additionally, using PV solar power to charge DC batteries, as used in Plug-in Hybrid Electric Vehicles and Electric Vehicles, leads to greater efficiencies. Traditionally, DC generated electricity from solar PV must be converted to AC for buildings, at an average 10% loss during the conversion. An additional efficiency loss occurs in the transition back to DC for battery driven devices and vehicles, and using various interest rates and energy price changes were calculated to find present values that range from $2,057.13 to $8,213.64 (analysis from 2009).

For example, in Germany with electricity prices of 0.25 euro/kWh and Insolation of 900 kWh/kW one kW_p will save 225 euro per year and with installation cost of 1700 euro/kW_p means that the system will pay back in less than 7 years.

Manufacturing

Overall the manufacturing process of creating solar photovoltaics is simple in that it does not require the culmination of many complex or moving parts. Because of the solid state nature of PV systems they often have relatively long lifetimes, anywhere from 10 to 30 years. In order to increase electrical output of a PV system the manufacturer must simply add more photovoltaic components and because of this economies of scale are important for manufacturers as costs decrease with increasing output.

While there are many types of PV systems known to be effective, crystalline silicon PV accounted for around 90% of the worldwide production of PV in 2013. Manufacturing silicon PV systems has several steps. First, polysilicon is processed from mined quartz until it is very pure (semi-conductor grade). This is melted down when small amounts of Boron, a group III element, are added to make a p-type semiconductor rich in electron holes. Typically using a seed crystal, an ingot of this solution is grown from the liquid polycrystalline. The ingot may also be cast in a mold. Wafers of this semiconductor material are cut from the bulk material with wire saws, and then go through surface etching before being cleaned. Next, the wafers are placed into a phosphorus vapor deposition furnace which lays a very thin layer of phosphorus, a group V element, which creates an N-type semiconducting surface. To reduce energy losses an anti-reflective coating is added to the surface, along with electrical contacts. After finishing the cell, cells are connected via electrical circuit according to the specific application and prepared for shipping and installation.

Crystalline silicon photovoltaics are only one type of PV, and while they represent the majority of solar cells produced currently there are many new and promising technologies that have the potential to be scaled up to meet future energy needs.

Another newer technology, thin-film PV, are manufactured by depositing semiconducting layers on substrate in vacuum. The substrate is often glass or stainless-steel, and these semiconducting layers are made of many types of materials including cadmium telluride (CdTe), copper indium diselenide (CIS), copper indium gallium diselenide (CIGS), and amorphous silicon (a-Si). After being deposited onto the substrate the semiconducting layers are separated and connected by electrical circuit by laser-scribing. Thin-film photovoltaics now make up around 20% of the overall production of PV because of the reduced materials requirements and cost to manufacture modules consisting of thin-films as compared to silicon-based wafers.

Other emerging PV technologies include organic, dye-sensitized, quantum-dot, and Perovskite photovoltaics. OPVs fall into the thin-film category of manufacturing, and typically operate around the 12% efficiency range which is lower than the 12–21% typically seen by silicon based PVs. Because organic photovoltaics require very high purity and are relatively reactive they must be encapsulated which vastly increases cost of manufacturing and meaning that they are not feasible for large scale up. Dye-sensitized PVs are similar in efficiency to OPVs but are significantly easier to manufacture. However these dye-sensitized photovoltaics present storage problems because the liquid electrolyte is toxic and can potentially permeate the plastics used in the cell. Quantum dot solar cells are quantum dot sensitized DSSCs and are solution processed meaning they are potentially scalable, but currently they have not reached greater than 10% efficiency. Perovskite solar cells are a very efficient solar energy converter and have excellent optoelectric properties for photovoltaic purposes, but they are expensive and difficult to manufacture.

Applications

Photovoltaic Systems

A photovoltaic system, or solar PV system is a power system designed to supply usable solar power by means of photovoltaics. It consists of an arrangement of several components, including solar panels to absorb and directly convert sunlight into electricity, a solar inverter to change the electric current from DC to AC, as well as mounting, cabling and other electrical accessories. PV systems range from small, roof-top mounted or building-integrated systems with capacities from a few to several tens of kilowatts, to large utility-scale power stations of hundreds of megawatts. Nowadays, most PV systems are grid-connected, while stand-alone systems only account for a small portion of the market.

- Rooftop and building integrated systems

Rooftop PV on half-timbered house

Photovoltaic arrays are often associated with buildings: either integrated into them, mounted on them or mounted nearby on the ground. Rooftop PV systems are most often retrofitted into existing buildings, usually mounted on top of the existing roof structure or on the existing walls. Alternatively, an array can be located separately from the building but connected by cable to supply power for the building. Building-integrated photovoltaics (BIPV) are increasingly incorporated into the roof or walls of new domestic and industrial buildings as a principal or ancillary source of electrical power. Roof tiles with integrated PV cells are sometimes used as well. Provided there is an open gap in which air can circulate, rooftop mounted solar panels can provide a passive cooling effect on buildings during the day and also keep accumulated heat in at night. Typically, residential rooftop systems have small capacities of around 5–10 kW, while commercial rooftop systems often amount to several hundreds of kilowatts. Although rooftop systems are much smaller than ground-mounted utility-scale power plants, they account for most of the worldwide installed capacity.

- Concentrator photovoltaics

Concentrator photovoltaics (CPV) is a photovoltaic technology that contrary to conventional flat-plate PV systems uses lenses and curved mirrors to focus sunlight onto small, but highly efficient, multi-junction (MJ) solar cells. In addition, CPV systems often use solar trackers and sometimes a cooling system to further increase their efficiency. Ongoing research and development is rapidly improving their competitiveness in the utility-scale segment and in areas of high solar insolation.

- Photovoltaic thermal hybrid solar collector

Photovoltaic thermal hybrid solar collector (PVT) are systems that convert solar radiation into thermal and electrical energy. These systems combine a solar PV cell, which converts sunlight into electricity, with a solar thermal collector, which captures the remaining energy and removes waste heat from the PV module. The capture of both electricity and heat allow these devices to have higher exergy and thus be more overall energy efficient than solar PV or solar thermal alone.

- Power stations

Many utility-scale solar farms have been constructed all over the world. As of 2015, the 579-megawatt (MW_{AC}) Solar Star is the world's largest photovoltaic power station, followed by the Desert Sunlight Solar Farm and the Topaz Solar Farm, both with a capacity of 550 MW_{AC}, constructed by US-company First Solar, using CdTe modules, a thin-film PV technology. All three power stations are located in the Californian desert. Many solar farms around the world are integrated with agriculture and some use innovative solar tracking systems that follow the sun's daily path across the sky to generate more electricity than conventional fixed-mounted systems. There are no fuel costs or emissions during operation of the power stations.

- Rural electrification

Developing countries where many villages are often more than five kilometers away from grid power are increasingly using photovoltaics. In remote locations in India a rural lighting program has been providing solar powered LED lighting to replace kerosene lamps. The solar powered lamps were sold at about the cost of a few months' supply of kerosene. Cuba is working to provide solar power for areas that are off grid. More complex applications of off-grid solar energy use include 3D printers. RepRap 3D printers have been solar powered with photovoltaic technology, which enables distributed manufacturing for sustainable development. These are areas where the social costs and benefits offer an excellent case for going solar, though the lack of profitability has relegated such endeavors to humanitarian efforts. However, in 1995 solar rural electrification projects had been found to be difficult to sustain due to unfavorable economics, lack of technical support, and a legacy of ulterior motives of north-to-south technology transfer.

- Standalone systems

Standalone PV system at an ecotourism resort (British Columbia, Canada).

Until a decade or so ago, PV was used frequently to power calculators and novelty devices. Improvements in integrated circuits and low power liquid crystal displays make it possible to power such devices for several years between battery changes, making PV use less common. In contrast, solar powered remote fixed devices have seen increasing use recently in locations where significant connection cost makes grid power prohibitively expensive. Such applications include solar lamps, water pumps, parking meters, emergency telephones, trash compactors, temporary traffic signs, charging stations, and remote guard posts and signals.

- Floatovoltaics

In May 2008, the Far Niente Winery in Oakville, CA pioneered the world's first "floatovoltaic" system by installing 994 photovoltaic solar panels onto 130 pontoons and floating them on the winery's irrigation pond. The floating system generates about 477 kW of peak output and when combined with an array of cells located adjacent to the pond is able to fully offset the winery's electricity consumption. The primary benefit of a floatovoltaic system is that it avoids the need to sacrifice valuable land area that could be used for another purpose. In the case of the Far Niente Winery, the floating system saved three-quarters of an acre that would have been required for a land-based system. That land area can instead be used for agriculture. Another benefit of a floatovoltaic system is that the panels are kept at a lower temperature than they would be on land, leading to a higher efficiency of solar energy conversion. The floating panels also reduce the amount of water lost through evaporation and inhibit the growth of algae.

- In transport

PV has traditionally been used for electric power in space. PV is rarely used to provide motive power in transport applications, but is being used increasingly to provide auxiliary power in boats and cars. Some automobiles are fitted with solar-powered air conditioning to limit interior temperatures on hot days. A self-contained solar vehicle would have limited power and utility, but a solar-charged electric vehicle allows use of solar power for transportation. Solar-powered cars, boats and airplanes have been demonstrated, with the most practical and likely of these being solar cars. The Swiss solar aircraft, Solar Impulse 2, achieved the longest non-stop solo flight in history and plan to make the first solar-powered aerial circumnavigation of the globe in 2015.

- Telecommunication and signaling

Solar PV power is ideally suited for telecommunication applications such as local telephone exchange, radio and TV broadcasting, microwave and other forms of electronic communication links. This is because, in most telecommunication application, storage batteries are already in use and the electrical system is basically DC. In hilly and mountainous terrain, radio and TV signals may not reach as they get blocked or reflected back due to undulating terrain. At these locations, low power transmitters (LPT) are installed to receive and retransmit the signal for local population.

- Spacecraft applications

Solar panels on spacecraft are usually the sole source of power to run the sensors, active heating and cooling, and communications. A battery stores this energy for use when the solar panels are in shadow. In some, the power is also used for spacecraft propulsion—electric propulsion. Spacecraft were one of the earliest applications of photovoltaics, starting with the silicon solar cells used on the Vanguard 1 satellite, launched by the US in 1958. Since then, solar power has been used on missions ranging from the MESSENGER probe to Mercury, to as far out in the solar system as the Juno probe to Jupiter. The largest solar power system flown in space is the electrical system of the International Space Station. To increase the power generated per kilogram, typical spacecraft solar panels use high-cost, high-efficiency, and close-packed rectangular multi-junction solar cells made of gallium arsenide (GaAs) and other semiconductor materials.

Part of Juno's solar array

• Specialty Power Systems

Photovoltaics may also be incorporated as energy conversion devices for objects at elevated temperatures and with preferable radiative emissivities such as heterogeneous combustors.

Advantages

The 122 PW of sunlight reaching the Earth's surface is plentiful—almost 10,000 times more than the 13 TW equivalent of average power consumed in 2005 by humans. This abundance leads to the suggestion that it will not be long before solar energy will become the world's primary energy source. Additionally, solar electric generation has the highest power density (global mean of 170 W/m²) among renewable energies.

Solar power is pollution-free during use. Production end-wastes and emissions are manageable using existing pollution controls. End-of-use recycling technologies are under development and policies are being produced that encourage recycling from producers.

PV installations can operate for 100 years or even more with little maintenance or intervention after their initial set-up, so after the initial capital cost of building any solar power plant, operating costs are extremely low compared to existing power technologies.

Grid-connected solar electricity can be used locally thus reducing transmission/distribution losses (transmission losses in the US were approximately 7.2% in 1995).

Compared to fossil and nuclear energy sources, very little research money has been invested in the development of solar cells, so there is considerable room for improvement. Nevertheless, experimental high efficiency solar cells already have efficiencies of over 40% in case of concentrating photovoltaic cells and efficiencies are rapidly rising while mass-production costs are rapidly falling.

In some states of the United States, much of the investment in a home-mounted system may be lost if the home-owner moves and the buyer puts less value on the system than the seller. The city of Berkeley developed an innovative financing method to remove this limitation, by adding a tax assessment that is transferred with the home to pay for the solar panels. Now known as PACE, Property Assessed Clean Energy, 30 U.S. states have duplicated this solution.

There is evidence, at least in California, that the presence of a home-mounted solar system can actually increase the value of a home. According to a paper published in April 2011 by the Ernest Orlando Lawrence Berkeley National Laboratory titled An Analysis of the Effects of Residential Photovoltaic Energy Systems on Home Sales Prices in California:

The research finds strong evidence that homes with PV systems in California have sold for a premium over comparable homes without PV systems. More specifically, estimates for average PV premiums range from approximately $3.9 to $6.4 per installed watt (DC) among a large number of different model specifications, with most models coalescing near $5.5/watt. That value corresponds to a premium of approximately $17,000 for a relatively new 3,100 watt PV system (the average size of PV systems in the study).

Concentrated Solar Power

Concentrated solar power (also called concentrating solar power, concentrated solar thermal, and CSP) systems generate solar power by using mirrors or lenses to concentrate a large area of sunlight, or solar thermal energy, onto a small area. Electricity is generated when the concentrated light is converted to heat, which drives a heat engine (usually a steam turbine) connected to an electrical power generator or powers a thermochemical reaction (experimental as of 2013).

The three towers of the Ivanpah Solar Power Facility.

CSP is being widely commercialized and the CSP market saw about 740 megawatt (MW) of generating capacity added between 2007 and the end of 2010. More than half of this (about

478 MW) was installed during 2010, bringing the global total to 1095 MW. Spain added 400 MW in 2010, taking the global lead with a total of 632 MW, while the US ended the year with 509 MW after adding 78 MW, including two fossil–CSP hybrid plants. The Middle East is also ramping up their plans to install CSP based projects and as a part of that Plan, Shams-I which was the largest CSP Project in the world has been installed in Abu Dhabi, by Masdar. The largest CSP project in the world until January 2016 is Noor in Morocco and global operational power stands at 4,705 MW.

There is considerable academic and commercial interest internationally in a new form of CSP, called STEM, for off-grid applications to produce 24-hour industrial scale power for mining sites and remote communities in Italy, other parts of Europe, Australia, Asia, North Africa and Latin America. STEM uses fluidized silica sand as a thermal storage and heat transfer medium for CSP systems. It has been developed by Salerno-based Magaldi Industries. The first commercial application of STEM was scheduled to take place in Sicily from 2015.

CSP growth is expected to continue at a fast pace. As of January 2014, Spain had a total capacity of 2,300 MW making this country the world leader in CSP. United States follows with 1,740 MW. Interest is also notable in North Africa and the Middle East, as well as India and China. In Italy, a handful of companies are trying to get authorization for 14 plants, totalling 392 MW, despite a strong local and political opposition. The global market has been dominated by parabolic-trough plants, which account for 90% of CSP plants.

CSP is not to be confused with concentrator photovoltaics (CPV). In CPV, the concentrated sunlight is converted directly to electricity via the photovoltaic effect.

Ivanpah in California is running 69 percent below advertised power output, and one Spanish company, Abengoa, that commercialized CSP both in the US and abroad, is teetering on the brink of bankruptcy. CSP technologies currently cannot compete on price with photovoltaics (solar panels), which have experienced huge growth in recent years due to falling prices of the panels. Another drawback of Ivanpah is that it lacks thermal energy storage, one of the big advantages that CSP has over PV and most other renewables, which require either large scale energy storage systems like pumped hydro or fast acting natural gas power plants to be there as backup for those times when they are not producing much energy.

History

A legend has it that Archimedes used a "burning glass" to concentrate sunlight on the invading Roman fleet and repel them from Syracuse. In 1973 a Greek scientist, Dr. Ioannis Sakkas, curious about whether Archimedes could really have destroyed the Roman fleet in 212 BC, lined up nearly 60 Greek sailors, each holding an oblong mirror tipped to catch the sun's rays and direct them at a tar-covered plywood silhouette 160 feet away. The ship caught fire after a few minutes; however, historians continue to doubt the Archimedes story.

In 1866, Auguste Mouchout used a parabolic trough to produce steam for the first solar steam engine. The first patent for a solar collector was obtained by the Italian Alessandro Battaglia in Genoa, Italy, in 1886. Over the following years, inventors such as John Ericsson and Frank Shuman developed concentrating solar-powered devices for irrigation, refrigeration, and locomotion. In 1913 Shuman finished a 55 HP parabolic solar thermal energy station in Maadi, Egypt for irrigation. The first solar-power system using a mirror dish was built by Dr. R.H. Goddard, who was already well known for his research on liquid-fueled rockets and wrote an article in 1929 in which he asserted that all the previous obstacles had been addressed.

Professor Giovanni Francia (1911–1980) designed and built the first concentrated-solar plant, which entered into operation in Sant'Ilario, near Genoa, Italy in 1968. This plant had the architecture of today's concentrated-solar plants with a solar receiver in the center of a field of solar collectors. The plant was able to produce 1 MW with superheated steam at 100 bar and 500 °C. The 10 MW Solar One power tower was developed in Southern California in 1981, but the parabolic-trough technology of the nearby Solar Energy Generating Systems (SEGS), begun in 1984, was more workable. The 354 MW SEGS is still the largest solar power plant in the world, and will remain so until the 390 MW Ivanpah power tower project reaches full power.

Current Technology

CSP is used to produce electricity (sometimes called solar thermoelectricity, usually generated through steam). Concentrated-solar technology systems use mirrors or lenses with tracking systems to focus a large area of sunlight onto a small area. The concentrated light is then used as heat or as a heat source for a conventional power plant (solar thermoelectricity). The solar concentrators used in CSP systems can often also be used to provide industrial process heating or cooling, such as in solar air-conditioning.

Concentrating technologies exist in five common forms, namely parabolic trough, enclosed trough, dish Stirlings, concentrating linear Fresnel reflector, and solar power tower. Although simple, these solar concentrators are quite far from the theoretical maximum concentration. For example, the parabolic-trough concentration gives about ⅓ of the theoretical maximum for the design acceptance angle, that is, for the same overall tolerances for the system. Approaching the theoretical maximum may be achieved by using more elaborate concentrators based on nonimaging optics.

Different types of concentrators produce different peak temperatures and correspondingly varying thermodynamic efficiencies, due to differences in the way that they track the sun and focus light. New innovations in CSP technology are leading systems to become more and more cost-effective.

Parabolic Trough

A parabolic trough consists of a linear parabolic reflector that concentrates light onto a receiver positioned along the reflector's focal line. The receiver is a tube positioned directly above the middle of the parabolic mirror and filled with a working fluid. The reflector follows the sun during the daylight hours by tracking along a single axis. A working fluid (e.g. molten salt) is heated to 150–350 °C (300–660 °F) as it flows through the receiver and is then used as a heat source for a power generation system. Trough systems are the most developed CSP technology. The Solar Energy Generating Systems (SEGS) plants in California, the world's first commercial parabolic trough plants, Acciona's Nevada Solar One near Boulder City, Nevada, and Andasol, Europe's first commercial parabolic trough plant are representative, along with Plataforma Solar de Almería's SSPS-DCS test facilities in Spain.

Parabolic trough at a plant near Harper Lake, California

Enclosed Trough

Enclosed trough systems are used to absorb heat rather than process heat. The design encapsulates the solar thermal system within a greenhouse-like glasshouse. The glasshouse creates a protected environment to withstand the elements that can negatively impact reliability and efficiency of the solar thermal system. Lightweight curved solar-reflecting mirrors are suspended from the ceiling of the glasshouse by wires. A single-axis tracking system positions the mirrors to retrieve the optimal amount of sunlight. The mirrors concentrate the sunlight and focus it on a network of stationary steel pipes, also suspended from the glasshouse structure. Water is carried throughout the length of the pipe, which is boiled to generate steam when intense solar radiation is applied. Sheltering the mirrors from the wind allows them to achieve higher temperature rates and prevents dust from building up on the mirrors.

GlassPoint Solar, the company that created the Enclosed Trough design, states its technology can produce heat for Enhanced Oil Recovery (EOR) for about $5 per million British thermal units in sunny regions, compared to between $10 and $12 for other conventional solar thermal technologies.

Solar Power Tower

A solar power tower consists of an array of dual-axis tracking reflectors (heliostats) that concentrate sunlight on a central receiver atop a tower; the receiver contains a fluid deposit, which can consist of sea water. The working fluid in the receiver is heated to 500–1000 °C (773–1,273 K (932–1,832 °F)) and then used as a heat source for a power generation or energy storage system. An advantage of the solar tower is the reflectors can be adjusted instead of the whole tower. Power-tower development is less advanced than trough systems, but they offer higher efficiency and better energy storage capability. The Solar Two in Daggett, California and the CESA-1 in Plataforma Solar de Almeria Almeria, Spain, are the most representative demonstration plants. The Planta Solar 10 (PS10) in Sanlucar la Mayor, Spain, is the first commercial utility-scale solar power tower in the world. The 377 MW Ivanpah Solar Power Facility, located in the Mojave Desert, is the largest CSP facility in the world, and uses three power towers. The National Solar Thermal Test Facility, NSTTF located in Albuquerque, NM, is an experimental solar thermal test facility with a heliostat field capable of producing 6 MW.

The PS10 solar power plant in Andalucía, Spain, concentrates sunlight from a field of heliostats onto a central solar power tower.

Fresnel Reflectors

Fresnel reflectors are made of many thin, flat mirror strips to concentrate sunlight onto tubes through which working fluid is pumped. Flat mirrors allow more reflective surface in the same amount of space as a parabolic reflector, thus capturing more of the available sunlight, and they are much cheaper than parabolic reflectors. Fresnel reflectors can be used in various size CSPs.

Dish Stirling

A dish Stirling or dish engine system consists of a stand-alone parabolic reflector that concentrates light onto a receiver positioned at the reflector's focal point. The reflector tracks the Sun along two axes. The working fluid in the receiver is heated to 250–700 °C (480–1,300 °F) and then used by a Stirling engine to generate power. Parabolic-dish systems provide high solar-to-electric efficiency (between 31% and 32%), and their modular nature provides scalability. The Stirling Energy Systems (SES), United Sun Systems (USS) and Science Applications International Corporation (SAIC) dishes at UNLV, and Australian National University's Big Dish in Canberra, Australia are representative of this technology. A world record for solar to electric efficiency was set at 31.25% by SES dishes at the National Solar Thermal Test Facility (NSTTF) in New Mexico on January 31, 2008, a cold, bright day. According to its developer, Ripasso Energy, a Swedish firm, in 2015 its Dish Sterling system being tested in the Kalahari Desert in South Africa showed 34% efficiency. The SES installation in Maricopa, Phoenix was the largest Stirling Dish power installation in the world until it was sold to United Sun Systems. Subsequently, larger parts of the installation have been moved to China as part of the huge energy demand.

Solar Thermal Enhanced Oil Recovery

Heat from the sun can be used to provide steam used to make heavy oil less viscous and easier to pump. Solar power tower and parabolic troughs can be used to provide the steam which is used directly so no generators are required and no electricity is produced. Solar thermal enhanced oil recovery can extend the life of oilfields with very thick oil which would not otherwise be economical to pump.

Deployment Around The World

Worldwide CSP Capacity Since 1984 in MW_p

National CSP capacities in 2014 (MW_p)		
Country	**Total**	**Added**
Spain	2,300	0
United States	1,634	+752
India	225	+175
United Arab Emirates	100	0
Algeria	25	0
Egypt	20	0
Morocco	20	0
Australia	12	0
China	10	0
Thailand	5	0
Source: REN21 Global Status Report, September 2015		

The commercial deployment of CSP plants started by 1984 in the US with the SEGS plants. The last SEGS plant was completed in 1990. From 1991 to 2005 no CSP plants were built anywhere in the world. Global installed CSP-capacity has increased nearly tenfold since 2004 and grew at an average of 50 percent per year during the last five years. In 2013, worldwide installed capacity increased by 36 percent or nearly 0.9 gigawatt (GW) to more than 3.4 GW. Spain and the United States remained the global leaders, while the number of countries with installed CSP were growing. There is a notable trend towards developing countries and regions with high solar radiation.

CSP is also increasingly competing with the cheaper photovoltaic solar power and with concentrator photovoltaics (CPV), a fast-growing technology that just like CSP is suited best for regions of high solar insolation. In addition, a novel solar CPV/CSP hybrid system has been proposed recently.

Worldwide Concentrated Solar Power (MW_p)													
Year	1984	1985	1989	1990	...	2006	2007	2008	2009	2010	2011	2012	2013	2014
Installed	14	60	200	80	0	1	74	55	179	307	629	803	872	925
Cumulative	14	74	274	354	354	355	429	484	663	969	1,598	2,553	3,425	4,400
Sources: REN21 · CSP-world.com · IRENA													

Efficiency

The conversion efficiency η of the incident solar radiation into mechanical work – without considering the ultimate conversion step into electricity by a power generator – depends on the thermal radiation properties of the solar receiver and on the heat engine (e.g. steam turbine). Solar irradiation is first converted into heat by the solar receiver with the efficiency $\eta_{Receiver}$ and subsequently the heat is converted into work by the heat engine with the efficiency η_{Carnot}, using Carnot's principle. For a solar receiver providing a heat source at temperature T_H and a heat sink at room temperature T^0, the overall conversion efficiency can be calculated as follows:

$$\eta = \eta_{Receiver} \cdot \eta_{Carnot}$$

$$\text{with } \eta_{Carnot} = 1 - \frac{T^0}{T_H}$$

$$\text{and } \eta_{Receiver} = \frac{Q_{absorbed} - Q_{lost}}{Q_{solar}}$$

where Q_{solar}, $Q_{absorbed}$, Q_{lost} are respectively the incoming solar flux and the fluxes absorbed and lost by the system solar receiver.

For a solar flux I (e.g. I = 1000 W/m²) concentrated C times with an efficiency η_{Optics} on the system solar receiver with a collecting area A and an absorptivity α :

$$Q_{solar} = \eta_{Optics} ICA,$$

$$Q_{absorbed} = \alpha Q_{solar},$$

For simplicity's sake, one can assume that the losses are only radiative ones (a fair assumption for high temperatures), thus for a reradiating area A and an emissivity ϵ applying the Stefan-Boltzmann law yields:

$$Q_{lost} = A\epsilon\sigma T_H^4$$

Simplifying these equations by considering perfect optics (η_{Optics} = 1), collecting and reradiating areas equal and maximum absorptivity and emissivity (α = 1, ϵ = 1) then substituting in the first equation gives

$$\eta = \left(1 - \frac{\sigma T_H^4}{IC}\right) \cdot \left(1 - \frac{T^0}{T_H}\right)$$

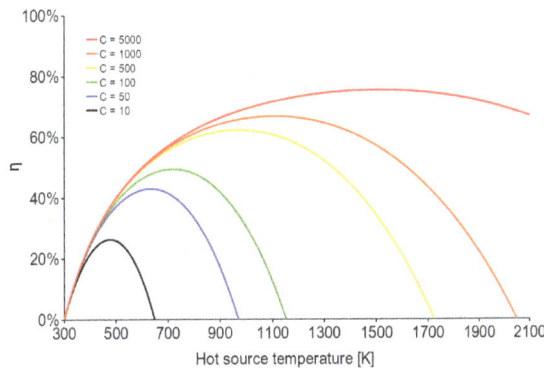

The graph shows that the overall efficiency does not increase steadily with the receiver's temperature. Although the heat engine's efficiency (Carnot) increases with higher temperature, the receiver's efficiency does not. On the contrary, the receiver's efficiency is decreasing, as the amount of energy it cannot absorb (Q_{lost}) grows by the fourth power as a function of temperature. Hence, there is a maximum reachable temperature. When the receiver efficiency is null (blue curve on the figure below), T_{max} is:

$$T_{max} = \left(\frac{IC}{\sigma}\right)^{0.25}$$

There is a temperature T_{opt} for which the efficiency is maximum, i.e. when the efficiency derivative relative to the receiver temperature is null:

$$\frac{d\eta}{dT_H}(T_{opt}) = 0$$

Consequently, this leads us to the following equation:

$$T_{opt}^5 - (0.75T^0)T_{opt}^4 - \frac{T^0 IC}{4\sigma} = 0$$

Solving this equation numerically allows us to obtain the optimum process temperature according to the solar concentration ratio C (red curve on the figure below)

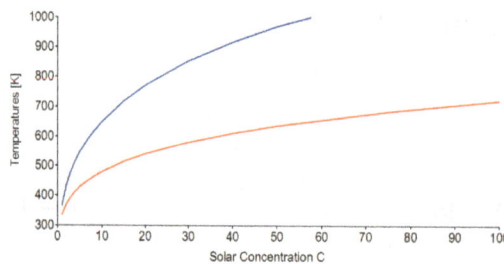

C	500	1000	5000	10000	45000 (MAX. FOR EARTH)
T_{MAX}	1720	2050	3060	3640	5300
T_{OPT}	970	1100	1500	1720	2310

Theoretical efficiencies aside, real-world experience of CSP reveals a 25-60 percent shortfall in projected production. A pilot 5 megawatt CSP power tower, Solar One, was converted to a 10 megawatt CSP power tower, Solar Two, decommissioned in 1999. Due to the success of Solar Two, a commercial power plant, called Solar Tres Power Tower, was built in Spain, renamed Gemasolar Thermosolar Plant. Gemasolar brilliant results have paved the way for the Crescent Dunes project. Ivanpah difficulties arise also from not having considered the lessons about the benefits of thermal storage. Solana in Arizona is 25 percent below projected numbers, Ivanpah in California, is 40 percent below projected numbers. A slightly bigger photovoltaic power station, like the 290 MW Agua

Caliente Solar Project peaked at most to 741 GW·h in 2014, comparing with the 280 MW Solana growing 719 GW·h. Lower insolation related to a solar minimum is part of the problem, but there is clearly a technical issue with rating these systems or simply a commercial overstatement. Indeed, an other operator, that of the 280 MW Genesis Solar, projected only 580 GW·h production and instead made 621 GW·h in 2015. CSP once thought to be economically superior to photovoltaics, in 2015 it has proven not to be the case. Recent PV commercial power is selling for ⅓ recent CSP contracts.

Costs

As of 9 September 2009, the cost of building a CSP station was typically about US$2.50 to $4 per watt, while the fuel (the sun's radiation) is free. Thus a 250 MW CSP station would have cost $600–1000 million to build. That works out to $0.12 to 0.18 USD/kWh. New CSP stations may be economically competitive with fossil fuels. Nathaniel Bullard, a solar analyst at Bloomberg New Energy Finance, has calculated that the cost of electricity at the Ivanpah Solar Power Facility, a project under construction in Southern California, will be lower than that from photovoltaic power and about the same as that from natural gas. However, in November 2011, Google announced that they would not invest further in CSP projects due to the rapid price decline of photovoltaics. Google invested US$168 million on BrightSource. IRENA has published on June 2012 a series of studies titled: "Renewable Energy Cost Analysis". The CSP study shows the cost of both building and operation of CSP plants. Costs are expected to decrease, but there are insufficient installations to clearly establish the learning curve. As of March 2012, there were 1.9 GW of CSP installed, with 1.8 GW of that being parabolic trough.

Incentives

Spain

Solar-thermal electricity generation is eligible for feed-in tariff payments (art. 2 RD 661/2007), if the system capacity does not exceed the following limits: Systems registered in the register of systems prior to 29 September 2008: 500 MW for solar-thermal systems. Systems registered after 29 September 2008 (PV only). The capacity limits for the different system types are re-defined during the review of the application conditions every quarter (art. 5 RD 1578/2008, Annex III RD 1578/2008). Prior to the end of an application period, the market caps specified for each system type are published on the website of the Ministry of Industry, Tourism and Trade (art. 5 RD 1578/2008).

Since 27 January 2012, Spain has halted acceptance of new projects for the feed-in-tariff. Projects currently accepted are not affected, except that a 6% tax on feed-in-tariffs has been adopted, effectively reducing the feed-in-tariff.

Australia

At the federal level, under the Large-scale Renewable Energy Target (LRET), in operation under the Renewable Energy Electricity Act 2000, large scale solar thermal electricity generation from accredited RET power stations may be entitled to create large-scale generation certificates (LGCs). These certificates can then be sold and transferred to liable entities (usually electricity retailers) to meet their obligations under this tradeable certificates scheme. However, as this legislation is technology neutral in its operation, it tends to favour more established RE technologies with a lower levelised cost of generation, such as large scale onshore wind, rather than solar thermal and CSP. At State level, renewable energy feed-in laws typically are capped by maximum generation capacity in kWp, and are open only to micro or medium scale generation and in a number of instances are only open to solar PV (photovoltaic) generation. This means that larger scale CSP projects would not be eligible for payment for feed-in incentives in many of the State and Territory jurisdictions.

Future

A study done by Greenpeace International, the European Solar Thermal Electricity Association, and the International Energy Agency's SolarPACES group investigated the potential and future of concentrated solar power. The study found that concentrated solar power could account for up to 25% of the world's energy needs by 2050. The increase in investment would be from 2 billion euros worldwide to 92.5 billion euros in that time period. Spain is the leader in concentrated solar power technology, with more than 50 government-approved projects in the works. Also, it exports its technology, further increasing the technology's stake in energy worldwide. Because the technology works best with areas of high insolation (solar radiation), experts predict the biggest growth in places like Africa, Mexico, and the southwest United States. It indicates that the thermal storage systems based in nitrates (calcium, potassium, sodium,...) will make the CSP plants more and more profitable. The study examined three different outcomes for this technology: no increases in CSP technology, investment continuing as it has been in Spain and the US, and finally the true potential of CSP without any barriers on its growth. The findings of the third part are shown in the table below:

Year	Annual Investment	Cumulative Capacity
2015	21 billion euros	420 megawatts
2050	174 billion euros	1,500,000 megawatts

Finally, the study acknowledged how technology for CSP was improving and how this would result in a drastic price decrease by 2050. It predicted a drop from the current range of €0.23–0.15/kwh to €0.14–0.10/kwh. Recently the EU has begun to look into developing a €400 billion ($774 billion) network of solar power plants based in the Sahara region using CSP technology known as Desertec, to create "a new carbon-free network linking Europe, the Middle East and North Africa". The plan is backed mainly by German industrialists and predicts production of 15% of Europe's power by 2050. Morocco is a major partner in Desertec and as it has barely 1% of the electricity consumption of the EU, it will produce more than enough energy for the entire country with a large energy surplus to deliver to Europe.

Algeria has the biggest area of desert, and private Algerian firm Cevital has signed up for Desertec. With its wide desert (the highest CSP potential in the Mediterranean and Middle East regions ~ about 170 TWh/year) and its strategic geographical location near Europe Algeria is one of the key countries to ensure the success of Desertec project. Moreover, with the abundant natural-gas reserve in the Algerian desert, this will strengthen the technical potential of Algeria in acquiring Solar-Gas Hybrid Power Plants for 24-hour electricity generation.

Other organizations expect CSP to cost $0.06(US)/kWh by 2015 due to efficiency improvements and mass production of equipment. That would make CSP as cheap as conventional power. Investors such as venture capitalist Vinod Khosla expect CSP to continuously reduce costs and actually be cheaper than coal power after 2015.

On 9 September 2009; 7 years ago, Bill Weihl, Google.org's green-energy spokesperson said that the firm was conducting research on the heliostat mirrors and gas turbine technology, which he expects will drop the cost of solar thermal electric power to less than $0.05/kWh in 2 or 3 years.

In 2009, scientists at the National Renewable Energy Laboratory (NREL) and SkyFuel teamed to develop large curved sheets of metal that have the potential to be 30% less expensive than today's best collectors of concentrated solar power by replacing glass-based models with a silver polymer sheet that has the same performance as the heavy glass mirrors, but at much lower cost and weight. It also is much easier to deploy and install. The glossy film uses several layers of polymers, with an inner layer of pure silver.

Telescope designer Roger Angel (Univ. of Arizona) has turned his attention to CPV, and is a partner in a company called Rehnu. Angel utilizes a spherical concentrating lens with large-telescope technologies, but much cheaper materials and mechanisms, to create efficient systems.

Recent experience with CSP technology in 2014 - 2015 at Solana in Arizona, and Ivanpah in Nevada indicate large production shortfalls in electricity generation between 25 and 40 percent. Producers blame clouds and stormy weather, but critics seem to think there are technological issues. These problems are causing utilities to pay inflated prices for wholesale electricity, and threaten the long-term viability of the technology. As photovoltaic costs continue to plummet, many think CSP has a limited future in utility-scale electricity production. On the other side, part of photovoltaic costs drop is the result of the fossil fuel cost reductions, which still powers most of PV production. Instead CSP would be unaffected when fossil prices return to higher quotes.

Very Large Scale Solar Power Plants

There are several proposals for gigawatt size, very large scale solar power plants. They include the Euro-Mediterranean Desertec proposal, Project Helios in Greece (10 gigawatt), and Ordos (2 gigawatt) in China. A 2003 study concluded that the world could generate 2,357,840 TWh each year from very large scale solar power plants using 1% of each of the world's deserts. Total consumption worldwide was 15,223 TWh/year (in 2003). The gigawatt size projects are arrays of single plants. The largest single plant in operation is the 370 MW Ivanpah Solar. In 2012, the BLM made available 97,921,069 acres (39,627,251 hectares) of land in the southwestern United States for solar projects, enough for between 10,000 and 20,000 gigawatts (GW).

Effect on Wildlife

It has been noted that insects can be attracted to the bright light caused by concentrated solar technology, and as a result birds that hunt them can be killed (burned) if the birds fly near the point where light is being focused onto. This can also affect raptors who hunt the birds. Federal wildlife officials have begun calling these power towers "mega traps" for wildlife.

However, the story about the Ivanpah Solar Power Facility was exaggerated, numbering the deaths in many tens of thousands, spreading alarm about concentrated solar power (CSP) plants, which was not grounded in facts, but on one opponent's speculation. According to rigorous reporting, in over six months, actually only 133 singed birds were counted. By focusing no more than 4 mirrors on any one place in the air during standby, at Crescent Dunes Solar Energy Project, in 3 months, the death rate dropped to zero fatalities.

References

- Solar Cells and their Applications Second Edition, Lewis Fraas, Larry Partain, Wiley, 2010, ISBN 978-0-470-44633-1 , Section10.2.

- Anderson, Lorraine; Palkovic, Rick (1994). Cooking with Sunshine (The Complete Guide to Solar Cuisine with 150 Easy Sun-Cooked Recipes). Marlowe & Company. ISBN 1-56924-300-X.

- Bradford, Travis (2006). Solar Revolution: The Economic Transformation of the Global Energy Industry. MIT Press. ISBN 0-262-02604-X.

- Butti, Ken; Perlin, John (1981). A Golden Thread (2500 Years of Solar Architecture and Technology). Van Nostrand Reinhold. ISBN 0-442-24005-8.

- Martin, Christopher L.; Goswami, D. Yogi (2005). Solar Energy Pocket Reference. International Solar Energy Society. ISBN 0-9771282-0-2.

- Scheer, Hermann (2002). The Solar Economy (Renewable Energy for a Sustainable Global Future). Earthscan Publications Ltd. ISBN 1-84407-075-1.

- Schittich, Christian (2003). Solar Architecture (Strategies Visions Concepts). Architektur-Dokumentation GmbH & Co. KG. ISBN 3-7643-0747-1.

- Smil, Vaclav (2006). Energy at the Crossroads (PDF). Organisation for Economic Co-operation and Development. ISBN 0-262-19492-9. Retrieved 29 September 2007.

- Palz, Wolfgang (2013). Solar Power for the World: What You Wanted to Know about Photovoltaics. CRC Press. pp. 131–. ISBN 978-981-4411-87-5.

- Ken Butti, John Perlin (1980) A Golden Thread: 2500 Years of Solar Architecture and Technology, Cheshire Books, pp. 66–100, ISBN 0442240058.

- "World's Largest CSP in Morocco Attracts South-South Learning | Climate Investment Funds". climateinvestmentfunds.org. Retrieved 2016-01-12.

- "Solare Termodinamico, una ricchezza per il paese, p.21" (PDF). Associazione Nazionale Energia Solare Termodinamica (in Italian). Retrieved 2016-02-07.

- Worland, Justin (4 April 2016). "After years of torrid growth, residential solar power faces serious growing pains". Time. Vol. 187 no. 12. p. 24. Retrieved 10 April 2016.

- "Electric cars not solar panels, says Environment Commissioner". Parliamentary Commissioner for the Environment. 22 March 2016. Retrieved 23 March 2016.

- Frank, Dimroth. "New world record for solar cell efficiency at 46% French-German cooperation confirms competitive advantage of European photovoltaic industry". Fraunhofer-Gesellschaft. Retrieved 14 March 2016.

5

Understanding Geothermal Energy

The thermal energy generated and stored in the earth is defined as geothermal energy. To have a better understanding on geothermal energy, this chapter elucidates concepts such as geothermal gradient, geothermal power and binary cycle. Geothermal energy is an emerging field of study, the following chapter will not only provide an overview, it will also delve deep into the variegated topics related to it.

Geothermal Energy

Geothermal energy is thermal energy generated and stored in the Earth. Thermal energy is the energy that determines the temperature of matter. The geothermal energy of the Earth's crust originates from the original formation of the planet and from radioactive decay of materials (in currently uncertain but possibly roughly equal proportions). The geothermal gradient, which is the difference in temperature between the core of the planet and its surface, drives a continuous conduction of thermal energy in the form of heat from the core to the surface.

Steam rising from the Nesjavellir Geothermal Power Station in Iceland.

Earth's internal heat is thermal energy generated from radioactive decay and continual heat loss from Earth's formation. Temperatures at the core–mantle boundary may reach over 4000 °C (7,200 °F). The high temperature and pressure in Earth's interior cause some rock to melt and solid mantle to behave plastically, resulting in portions of mantle convecting upward since it is lighter than the surrounding rock. Rock and water is heated in the crust, sometimes up to 370 °C (700 °F).

From hot springs, geothermal energy has been used for bathing since Paleolithic times and for space heating since ancient Roman times, but it is now better known for electricity generation. Worldwide, 11,700 megawatts (MW) of geothermal power is online in 2013. An additional 28 gigawatts of direct geothermal heating capacity is installed for district heating, space heating, spas, industrial processes, desalination and agricultural applications in 2010.

Geothermal power is cost-effective, reliable, sustainable, and environmentally friendly, but has historically been limited to areas near tectonic plate boundaries. Recent technological advances have dramatically expanded the range and size of viable resources, especially for applications such as home heating, opening a potential for widespread exploitation. Geothermal wells release greenhouse gases trapped deep within the earth, but these emissions are much lower per energy unit than those of fossil fuels. As a result, geothermal power has the potential to help mitigate global warming if widely deployed in place of fossil fuels.

The Earth's geothermal resources are theoretically more than adequate to supply humanity's energy needs, but only a very small fraction may be profitably exploited. Drilling and exploration for deep resources is very expensive. Forecasts for the future of geothermal power depend on assumptions about technology, energy prices, subsidies, and interest rates. Pilot programs like EWEB's customer opt in Green Power Program show that customers would be willing to pay a little more for a renewable energy source like geothermal. But as a result of government assisted research and industry experience, the cost of generating geothermal power has decreased by 25% over the past two decades. In 2001, geothermal energy costs between two and ten US cents per kWh.

History

The oldest known pool fed by a hot spring, built in the Qin dynasty in the 3rd century BCE.

Hot springs have been used for bathing at least since Paleolithic times. The oldest known spa is a stone pool on China's Lisan mountain built in the Qin Dynasty in the 3rd century BC, at the same site where the Huaqing Chi palace was later built. In the first century AD, Romans conquered *Aquae Sulis*, now Bath, Somerset, England, and used the hot springs there to feed public baths and underfloor heating. The admission fees for these baths probably represent the first commercial use of geothermal power. The world's oldest geothermal district heating system in Chaudes-Aigues,

France, has been operating since the 14th century. The earliest industrial exploitation began in 1827 with the use of geyser steam to extract boric acid from volcanic mud in Larderello, Italy.

In 1892, America's first district heating system in Boise, Idaho was powered directly by geothermal energy, and was copied in Klamath Falls, Oregon in 1900. A deep geothermal well was used to heat greenhouses in Boise in 1926, and geysers were used to heat greenhouses in Iceland and Tuscany at about the same time. Charlie Lieb developed the first downhole heat exchanger in 1930 to heat his house. Steam and hot water from geysers began heating homes in Iceland starting in 1943.

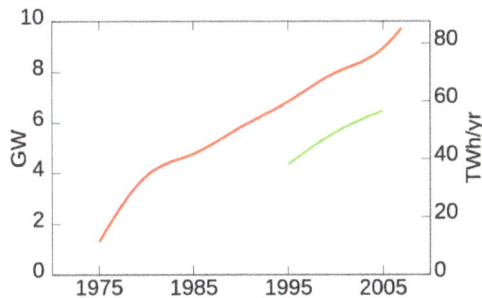

Global geothermal electric capacity. Upper red line is installed capacity; lower green line is realized production.

In the 20th century, demand for electricity led to the consideration of geothermal power as a generating source. Prince Piero Ginori Conti tested the first geothermal power generator on 4 July 1904, at the same Larderello dry steam field where geothermal acid extraction began. It successfully lit four light bulbs. Later, in 1911, the world's first commercial geothermal power plant was built there. It was the world's only industrial producer of geothermal electricity until New Zealand built a plant in 1958. In 2012, it produced some 594 megawatts.

Lord Kelvin invented the heat pump in 1852, and Heinrich Zoelly had patented the idea of using it to draw heat from the ground in 1912. But it was not until the late 1940s that the geothermal heat pump was successfully implemented. The earliest one was probably Robert C. Webber's home-made 2.2 kW direct-exchange system, but sources disagree as to the exact timeline of his invention. J. Donald Kroeker designed the first commercial geothermal heat pump to heat the Commonwealth Building (Portland, Oregon) and demonstrated it in 1946. Professor Carl Nielsen of Ohio State University built the first residential open loop version in his home in 1948. The technology became popular in Sweden as a result of the 1973 oil crisis, and has been growing slowly in worldwide acceptance since then. The 1979 development of polybutylene pipe greatly augmented the heat pump's economic viability.

In 1960, Pacific Gas and Electric began operation of the first successful geothermal electric power plant in the United States at The Geysers in California. The original turbine lasted for more than 30 years and produced 11 MW net power.

The binary cycle power plant was first demonstrated in 1967 in the USSR and later introduced to the US in 1981. This technology allows the generation of electricity from much lower temperature resources than previously. In 2006, a binary cycle plant in Chena Hot Springs, Alaska, came online, producing electricity from a record low fluid temperature of 57 °C (135 °F).

Direct Usage

Direct Use Data 2015	
Country	**Usage (MWt) 2015**
United States	17,415.91
Philippines	3.30
Indonesia	2.30
Mexico	155.82
Italy	1,014.00
New Zealand	487.45
Iceland	2,040.00
Japan	2,186.17
Iran	81.50
El Salvador	3.36
Kenya	22.40
Costa Rica	1.00
Russia	308.20
Turkey	2,886.30
Papua-New Guinea	0.10
Guatemala	2.31
Portugal	35.20
China	17,870.00
France	2,346.90
Ethiopia	2.20
Germany	2,848.60
Austria	903.40
Australia	16.09
Thailand	128.51

Electricity

The International Geothermal Association (IGA) has reported that 10,715 megawatts (MW) of geothermal power in 24 countries is online, which was expected to generate 67,246 GWh of electricity in 2010. This represents a 20% increase in online capacity since 2005. IGA projects growth to 18,500 MW by 2015, due to the projects presently under consideration, often in areas previously assumed to have little exploitable resource.

In 2010, the United States led the world in geothermal electricity production with 3,086 MW of installed capacity from 77 power plants. The largest group of geothermal power plants in the world is located at The Geysers, a geothermal field in California. The Philippines is the second highest producer, with 1,904 MW of capacity online. Geothermal power makes up approximately 27% of Philippine electricity generation.

Installed geothermal electric capacity				
Country	Capacity (MW) 2007	Capacity (MW) 2010	Percentage of national electricity production	Percentage of global geothermal production
United States	2687	3086	0.3	29
Philippines	1969.7	1904	27	18
Indonesia	992	1197	3.7	11
Mexico	953	958	3	9
Italy	810.5	843	1.5	8
New Zealand	471.6	628	10	6
Iceland	421.2	575	30	5
Japan	535.2	536	0.1	5
Iran	250	250		
El Salvador	204.2	204	25	
Kenya	128.8	167	11.2	
Costa Rica	162.5	166	14	
Nicaragua	87.4	88	10	
Russia	79	82		
Turkey	38	82		
Papua-New Guinea	56	56		
Guatemala	53	52		
Portugal	23	29		
China	27.8	24		
France	14.7	16		
Ethiopia	7.3	7.3		
Germany	8.4	6.6		
Austria	1.1	1.4		
Australia	0.2	1.1		
Thailand	0.3	0.3		
TOTAL	9,981.9	10,959.7		

Geothermal electric plants were traditionally built exclusively on the edges of tectonic plates where high temperature geothermal resources are available near the surface. The development of binary cycle power plants and improvements in drilling and extraction technology enable enhanced geothermal systems over a much greater geographical range. Demonstration projects are operational in Landau-Pfalz, Germany, and Soultz-sous-Forêts, France, while an earlier effort in Basel, Switzerland was shut down after it triggered earthquakes. Other demonstration projects are under construction in Australia, the United Kingdom, and the United States of America.

The thermal efficiency of geothermal electric plants is low, around 10–23%, because geothermal fluids do not reach the high temperatures of steam from boilers. The laws of thermodynamics limits the efficiency of heat engines in extracting useful energy. Exhaust heat is wasted, unless it can be used directly and locally, for example in greenhouses, timber mills, and district heating. System efficiency does not materially affect operational costs as it would for plants that use fuel, but it does affect return on the capital used to build the plant. In order to produce more energy than the pumps consume, electricity generation requires relatively hot fields and specialized heat cycles. Because geothermal power does not rely on variable sources of energy, unlike, for example, wind or solar, its capacity factor can be quite large – up to 96% has been demonstrated. The global average was 73% in 2005.

Types

Geothermal energy comes in either *vapor-dominated* or *liquid-dominated* forms. Larderello and The Geysers are vapor-dominated. Vapor-dominated sites offer temperatures from 240 to 300 °C that produce superheated steam.

Liquid-dominated Plants

Liquid-dominated reservoirs (LDRs) were more common with temperatures greater than 200 °C (392 °F) and are found near young volcanoes surrounding the Pacific Ocean and in rift zones and hot spots. *Flash plants* are the common way to generate electricity from these sources. Pumps are generally not required, powered instead when the water turns to steam. Most wells generate 2-10MWe. Steam is separated from liquid via cyclone separators, while the liquid is returned to the reservoir for reheating/reuse. As of 2013, the largest liquid system is Cerro Prieto in Mexico, which generates 750 MWe from temperatures reaching 350 °C (662 °F). The Salton Sea field in Southern California offers the potential of generating 2000 MWe.

Lower temperature LDRs (120–200 °C) require pumping. They are common in extensional terrains, where heating takes place via deep circulation along faults, such as in the Western US and Turkey. Water passes through a heat exchanger in a Rankine cycle binary plant. The water vaporizes an organic working fluid that drives a turbine. These binary plants originated in the Soviet Union in the late 1960s and predominate in new US plants. Binary plants have no emissions.

Thermal Energy

Lower temperature sources produce the energy equivalent of 100M BBL per year. Sources with temperatures of 30–150 °C are used without conversion to electricity as district heating, greenhouses, fisheries, mineral recovery, industrial process heating and bathing in 75 countries. Heat pumps extract energy from shallow sources at 10–20 °C in 43 countries for use in space heating and cooling. Home heating is the fastest-growing means of exploiting geothermal energy, with global annual growth rate of 30% in 2005 and 20% in 2012.

Approximately 270 petajoules (PJ) of geothermal heating was used in 2004. More than half went for space heating, and another third for heated pools. The remainder supported industrial and agricultural applications. Global installed capacity was 28 GW, but capacity factors tend to be low (30% on average) since heat is mostly needed in winter. Some 88 PJ for space heating was extracted by an estimated 1.3 million geothermal heat pumps with a total capacity of 15 GW.

Heat for these purposes may also be extracted from co-generation at a geothermal electrical plant.

Heating is cost-effective at many more sites than electricity generation. At natural hot springs or geysers, water can be piped directly into radiators. In hot, dry ground, earth tubes or downhole heat exchangers can collect the heat. However, even in areas where the ground is colder than room temperature, heat can often be extracted with a geothermal heat pump more cost-effectively and cleanly than by conventional furnaces. These devices draw on much shallower and colder resources than traditional geothermal techniques. They frequently combine functions, including air conditioning, seasonal thermal energy storage, solar energy collection, and electric heating. Heat pumps can be used for space heating essentially anywhere.

Iceland is the world leader in direct applications. Some 92.5% of its homes are heated with geothermal energy, saving Iceland over $100 million annually in avoided oil imports. Reykjavík, Iceland has the world's biggest district heating system. Once known as the most polluted city in the world, it is now one of the cleanest.

Enhanced Geothermal

Enhanced geothermal systems (EGS) actively inject water into wells to be heated and pumped back out. The water is injected under high pressure to expand existing rock fissures to enable the water to freely flow in and out. The technique was adapted from oil and gas extraction techniques. However, the geologic formations are deeper and no toxic chemicals are used, reducing the possibility of environmental damage. Drillers can employ directional drilling to expand the size of the reservoir.

Small-scale EGS have been installed in the Rhine Graben at Soultz-sous-Forêts in France and at Landau and Insheim in Germany.

Economics

Geothermal power requires no fuel (except for pumps), and is therefore immune to fuel cost fluctuations. However, capital costs are significant. Drilling accounts for over half the costs, and exploration of deep resources entails significant risks. A typical well doublet (extraction and injection wells) in Nevada can support 4.5 megawatts (MW) and costs about $10 million to drill, with a 20% failure rate.

A power plant at The Geysers

In total, electrical plant construction and well drilling cost about €2–5 million per MW of electrical capacity, while the break–even price is 0.04–0.10 € per kW·h. Enhanced geothermal systems tend to be on the high side of these ranges, with capital costs above $4 million per MW and break–even above \$0.054 per kW·h in 2007. Direct heating applications can use much shallower wells with lower temperatures, so smaller systems with lower costs and risks are feasible. Residential geothermal heat pumps with a capacity of 10 kilowatt (kW) are routinely installed for around \$1–3,000 per kilowatt. District heating systems may benefit from economies of scale if demand is geographically dense, as in cities and greenhouses, but otherwise piping installation dominates capital costs. The capital cost of one such district heating system in Bavaria was estimated at somewhat over 1 million € per MW. Direct systems of any size are much simpler than electric generators and have lower maintenance costs per kW·h, but they must consume electricity to run pumps and compressors. Some governments subsidize geothermal projects.

Geothermal power is highly scalable: from a rural village to an entire city.

The most developed geothermal field in the United States is The Geysers in Northern California.

Geothermal projects have several stages of development. Each phase has associated risks. At the early stages of reconnaissance and geophysical surveys, many projects are cancelled, making that phase unsuitable for traditional lending. Projects moving forward from the identification, exploration and exploratory drilling often trade equity for financing.

Resources

The Earth's internal thermal energy flows to the surface by conduction at a rate of 44.2 terawatts (TW), and is replenished by radioactive decay of minerals at a rate of 30 TW. These power rates are more than double humanity's current energy consumption from all primary sources, but most of this energy flow is not recoverable. In addition to the internal heat flows, the top layer of the surface to a depth of 10 meters (33 ft) is heated by solar energy during the summer, and releases that energy and cools during the winter.

Outside of the seasonal variations, the geothermal gradient of temperatures through the crust is 25–30 °C (77–86 °F) per kilometer of depth in most of the world. The conductive heat flux averages 0.1 MW/km^2. These values are much higher near tectonic plate boundaries where the crust is thinner. They may be further augmented by fluid circulation, either through magma conduits, hot springs, hydrothermal circulation or a combination of these.

A geothermal heat pump can extract enough heat from shallow ground anywhere in the world to provide home heating, but industrial applications need the higher temperatures of deep resources. The thermal efficiency and profitability of electricity generation is particularly sensitive to temperature. The most demanding applications receive the greatest benefit from a high natural heat flux, ideally from using a hot spring. The next best option is to drill a well into a hot aquifer. If no adequate aquifer is available, an artificial one may be built by injecting water to hydraulically fracture the bedrock. This last approach is called hot dry rock geothermal energy in Europe, or enhanced geothermal systems in North America. Much greater potential may be available from this approach than from conventional tapping of natural aquifers.

Estimates of the potential for electricity generation from geothermal energy vary sixfold, from .035to2TW depending on the scale of investments. Upper estimates of geothermal resources assume enhanced geothermal wells as deep as 10 kilometres (6 mi), whereas existing geothermal wells are rarely more than 3 kilometres (2 mi) deep. Wells of this depth are now common in the petroleum industry. The deepest research well in the world, the Kola superdeep borehole, is 12 kilometres (7 mi) deep.

Production

According to the Geothermal Energy Association (GEA) installed geothermal capacity in the United States grew by 5%, or 147.05 MW, ce the last annual survey in March 2012. This increase came from seven geothermal projects that began production in 2012. GEA also revised its 2011 estimate of installed capacity upward by 128 MW, bringing current installed U.S. geothermal capacity to 3,386 MW.

Myanmar Engineering Society has identified at least 39 locations capable of geothermal power production and some of these hydrothermal reservoirs lie quite close to Yangon which is a significant underutilized resource.

Renewability and Sustainability

Geothermal power is considered to be renewable because any projected heat extraction is small compared to the Earth's heat content. The Earth has an internal heat content of 10^{31} joules ($3 \cdot 10^{15}$ TW·hr), approximately 100 billion times current (2010) worldwide annual energy consumption. About 20% of this is residual heat from planetary accretion, and the remainder is attributed to higher radioactive decay rates that existed in the past. Natural heat flows are not in equilibrium, and the planet is slowly cooling down on geologic timescales. Human extraction taps a minute fraction of the natural outflow, often without accelerating it.

Geothermal power is also considered to be sustainable thanks to its power to sustain the Earth's intricate ecosystems. By using geothermal sources of energy present generations of humans will not endanger the capability of future generations to use their own resources to the same amount that those energy sources are presently used. Further, due to its low emissions geothermal energy is considered to have excellent potential for mitigation of global warming.

Even though geothermal power is globally sustainable, extraction must still be monitored to avoid local depletion. Over the course of decades, individual wells draw down local temperatures and water levels until a new equilibrium is reached with natural flows. The three oldest sites, at Larderello, Wairakei, and the Geysers have experienced reduced output because of local depletion. Heat and water, in uncertain proportions, were extracted faster than they were replenished. If production is reduced and water is reinjected, these wells could theoretically recover their full potential. Such mitigation strategies have already been implemented at some sites. The long-term sustainability of geothermal energy has been demonstrated at the Lardarello field in Italy since 1913, at the Wairakei field in New Zealand since 1958, and at The Geysers field in California since 1960.

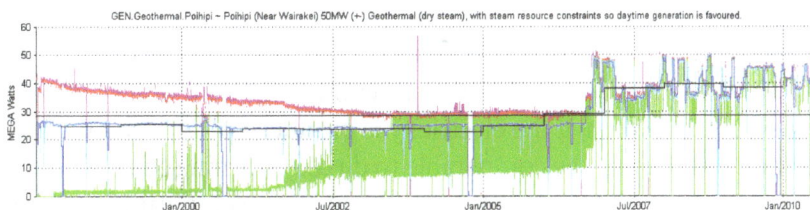

Electricity Generation at Poihipi, New Zealand.

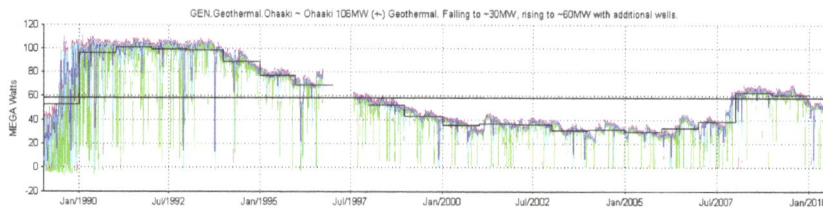

Electricity Generation at Ohaaki, New Zealand.

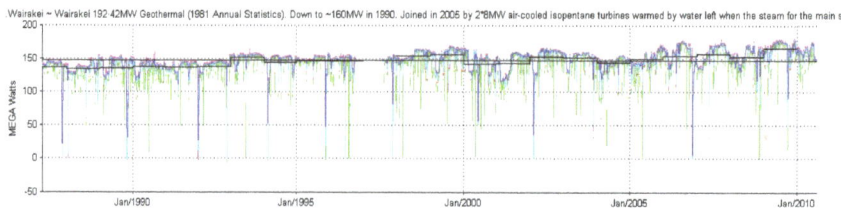

Electricity Generation at Wairakei, New Zealand.

Falling electricity production may be boosted through drilling additional supply boreholes, as at Poihipi and Ohaaki. The Wairakei power station has been running much longer, with its first unit commissioned in November 1958, and it attained its peak generation of 173MW in 1965, but already the supply of high-pressure steam was faltering, in 1982 being derated to intermediate pressure and the station managing 157MW. Around the start of the 21st century it was managing about 150MW, then in 2005 two 8MW isopentane systems were added, boosting the station's output by about 14MW. Detailed data are unavailable, being lost due to re-organisations. One such re-organisation in 1996 causes the absence of early data for Poihipi (started 1996), and the gap in 1996/7 for Wairakei and Ohaaki; half-hourly data for Ohaaki's first few months of operation are also missing, as well as for most of Wairakei's history.

Environmental Effects

Fluids drawn from the deep earth carry a mixture of gases, notably carbon dioxide (CO_2), hydrogen sulfide (H_2S), methane (CH_4) and ammonia (NH_3). These pollutants contribute to global warming, acid rain, and noxious smells if released. Existing geothermal electric plants emit an average of 122 kilograms (269 lb) of CO_2 per megawatt-hour (MW·h) of electricity, a small fraction of the emission intensity of conventional fossil fuel plants. Plants that experience high levels of acids and volatile chemicals are usually equipped with emission-control systems to reduce the exhaust.

In addition to dissolved gases, hot water from geothermal sources may hold in solution trace amounts of toxic elements such as mercury, arsenic, boron, and antimony. These chemicals precipitate as the water cools, and can cause environmental damage if released. The modern practice of injecting cooled geothermal fluids back into the Earth to stimulate production has the side benefit of reducing this environmental risk.

Direct geothermal heating systems contain pumps and compressors, which may consume energy from a polluting source. This parasitic load is normally a fraction of the heat output, so it is always less polluting than electric heating. However, if the electricity is produced by burning fossil fuels, then the net emissions of geothermal heating may be comparable to directly burning the fuel for heat. For example, a geothermal heat pump powered by electricity from a combined cycle natural gas plant would produce about as much pollution as a natural gas condensing furnace of the same size. Therefore, the environmental value of direct geothermal heating applications is highly dependent on the emissions intensity of the neighboring electric grid.

Plant construction can adversely affect land stability. Subsidence has occurred in the Wairakei field in New Zealand. In Staufen im Breisgau, Germany, tectonic uplift occurred instead, due to a previously isolated anhydrite layer coming in contact with water and turning into gypsum, doubling its volume. Enhanced geothermal systems can trigger earthquakes as part of hydraulic fracturing. The project in Basel, Switzerland was suspended because more than 10,000 seismic events measuring up to 3.4 on the Richter Scale occurred over the first 6 days of water injection.

Geothermal has minimal land and freshwater requirements. Geothermal plants use 3.5 square kilometres (1.4 sq mi) per gigawatt of electrical production (not capacity) versus 32 square kilometres (12 sq mi) and 12 square kilometres (4.6 sq mi) for coal facilities and wind farms respectively. They use 20 litres (5.3 US gal) of freshwater per MW·h versus over 1,000 litres (260 US gal) per MW·h for nuclear, coal, or oil.

Legal Frameworks

Some of the legal issues raised by geothermal energy resources include questions of ownership and allocation of the resource, the grant of exploration permits, exploitation rights, royalties, and the extent to which geothermal energy issues have been recognized in existing planning and environmental laws. Other questions concern overlap between geothermal and mineral or petroleum tenements. Broader issues concern the extent to which the legal framework for encouragement of renewable energy assists in encouraging geothermal industry innovation and development.

Geothermal Gradient

Geothermal gradient is the rate of increasing temperature with respect to increasing depth in the Earth's interior. Away from tectonic plate boundaries, it is about 25 °C per km of depth (1 °F per 70 feet of depth) near the surface in most of the world. Strictly speaking, *geo*-thermal necessarily refers to the Earth but the concept may be applied to other planets. A line tracing the gradient through the planetary body is called a geotherm on Earth and other terrestrial planets. On the Moon it is called a selenotherm.

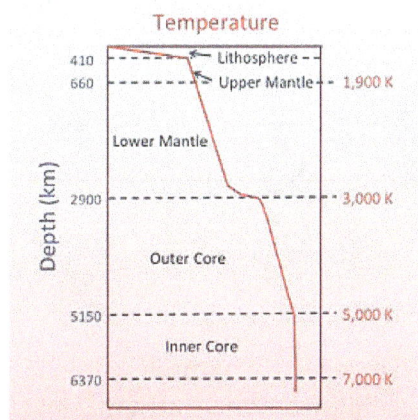

Temperature profile of the inner Earth

The Earth's internal heat comes from a combination of residual heat from planetary accretion, heat produced through radioactive decay, and possibly heat from other sources. The major heat-producing isotopes in the Earth are potassium-40, uranium-238, uranium-235, and thorium-232. At the center of the planet, the temperature may be up to 7,000 K and the pressure could reach 360 GPa(3.6 million atm). Because much of the heat is provided by radioactive decay, scientists believe that early in Earth history, before isotopes with short half-lives had been depleted, Earth's heat production would have been much higher. Heat production was twice that of present-day at approximately 3 billion years ago, resulting in larger temperature gradients within the Earth, larger rates of mantle convection and plate tectonics, allowing the production of igneous rocks such as komatiites that are not formed anymore today.

Heat Sources

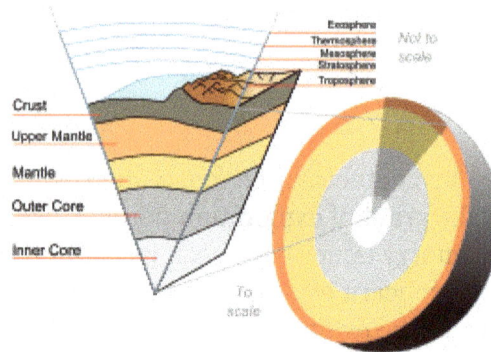

Earth cutaway from core to exosphere

Temperature within the Earth increases with depth. Highly viscous or partially molten rock at temperatures between 650 to 1,200 °C (1,200 to 2,200 °F) are found at the margins of tectonic plates, increasing the geothermal gradient in the vicinity, but only the outer core is postulated to exist in a molten or fluid state, and the temperature at the Earth's inner core/outer core boundary, around 3,500 kilometres (2,200 mi) deep, is estimated to be 5650 ± 600 kelvins. The heat content of the Earth is 10^{31} joules.

Geothermal drill machine in Wisconsin, USA

- Much of the heat is created by decay of naturally radioactive elements. An estimated 45 to 90 percent of the heat escaping from the Earth originates from radioactive decay of elements mainly located in the mantle.

- Heat of impact and compression released during the original formation of the Earth by accretion of in-falling meteorites.

- Heat released as abundant heavy metals (iron, nickel, copper) descended to the Earth's core.

- Latent heat released as the liquid outer core crystallizes at the inner core boundary.

- Heat may be generated by tidal force on the Earth as it rotates; since rock cannot flow as readily as water it compresses and distorts, generating heat.

- There is no reputable science to suggest that any significant heat may be created by electromagnetic effects of the magnetic fields involved in Earth's magnetic field, as suggested by some contemporary folk theories.

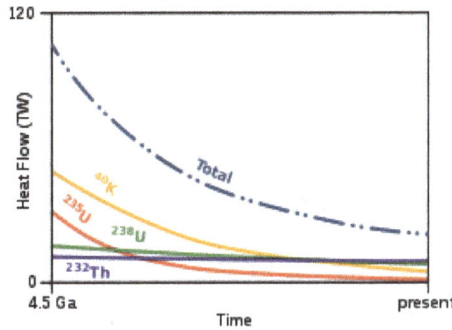

The radiogenic heat from the decay of 238U and 232Th are now the major contributors to the earth's internal heat budget.

In Earth's continental crust, the decay of natural radioactive isotopes has had significant involvement in the origin of geothermal heat. The continental crust is abundant in lower density minerals but also contains significant concentrations of heavier lithophilic minerals such as uranium. Because of this, it holds the largest global reservoir of radioactive elements found in the Earth. Especially in layers closer to Earth's surface, naturally occurring isotopes are enriched in the granite and basaltic rocks. These high levels of radioactive elements are largely excluded from the Earth's mantle due to their inability to substitute in mantle minerals and consequent enrichment in partial melts. The mantle is mostly made up of high density minerals with high contents of atoms that have relatively small atomic radii such as magnesium (Mg), titanium (Ti), and calcium (Ca).

Present-day major heat-producing isotopes				
Isotope	Heat release [W/kg isotope]	Half-life [years]	Mean mantle concentration [kg isotope/kg mantle]	Heat release [W/kg mantle]
^{238}U	9.46×10^{-5}	4.47×10^{9}	30.8×10^{-9}	2.91×10^{-12}
^{235}U	5.69×10^{-4}	7.04×10^{8}	0.22×10^{-9}	1.25×10^{-13}
^{232}Th	2.64×10^{-5}	1.40×10^{10}	124×10^{-9}	3.27×10^{-12}
^{40}K	2.92×10^{-5}	1.25×10^{9}	36.9×10^{-9}	1.08×10^{-12}

Heat Flow

Heat flows constantly from its sources within the Earth to the surface. Total heat loss from the Earth is estimated at 44.2 TW (4.42×10^{13} watts). Mean heat flow is 65 mW/m² over continental crust and 101 mW/m² over oceanic crust. This is 0.087 watt/square meter on average (0.03 percent of solar power absorbed by the Earth), but is much more concentrated in areas where thermal energy is transported toward the crust by convection such as along mid-ocean ridges and mantle plumes. The Earth's crust effectively acts as a thick insulating blanket which must be pierced by fluid conduits (of magma, water or other) in order to release the heat underneath. More of the heat in the Earth is lost through plate tectonics, by mantle upwelling associated with mid-ocean ridges. The final major mode of heat loss is by conduction through the lithosphere, the majority of which occurs in the oceans due to the crust there being much thinner and younger than under the continents.

The heat of the Earth is replenished by radioactive decay at a rate of 30 TW. The global geothermal flow rates are more than twice the rate of human energy consumption from all primary sources.

Direct Application

Heat from Earth's interior can be used as an energy source, known as geothermal energy. The geothermal gradient has been used for space heating and bathing since ancient Roman times, and more recently for generating electricity. As the human population continues to grow, so does energy use and the correlating environmental impacts that are consistent with global primary sources of energy. This has caused a growing interest in finding sources of energy that are renewable and have reduced greenhouse gas emissions. In areas of high geothermal energy density, current technology allows for the generation of electrical power because of the corresponding high temperatures. Generating electrical power from geothermal resources requires no fuel while providing true baseload energy at a reliability rate that constantly exceeds 90%. In order to extract geothermal energy, it is necessary to efficiently transfer heat from a geothermal reservoir to a power plant, where electrical energy is converted from heat. On a worldwide scale, the heat stored in Earth's interior provides an energy that is still seen as an exotic source. About 10 GW of geothermal electric capacity is installed around the world as of 2007, generating 0.3% of global electricity demand. An additional 28 GW of direct geothermal heating capacity is installed for district heating, space heating, spas, industrial processes, desalination and agricultural applications. Because heat is flowing through every square meter of land, it can be used for a source of energy for heating, air conditioning (HVAC) and ventilating systems using ground source heat pumps. In areas where modest heat flow is present, geothermal energy can be used for industrial applications that presently rely on fossil fuels.

Variations

The geothermal gradient varies with location and is typically measured by determining the bottom open-hole temperature after borehole drilling. To achieve accuracy the drilling fluid needs time to reach the ambient temperature. This is not always achievable for practical reasons.

In stable tectonic areas in the tropics a temperature-depth plot will converge to the annual average surface temperature. However, in areas where deep permafrost developed during the Pleistocene a low temperature anomaly can be observed that persists down to several hundred metres. The Suwałki cold anomaly in Poland has led to the recognition that similar thermal disturbances related to Pleistocene-Holocene climatic changes are recorded in boreholes throughout Poland, as well as in Alaska, northern Canada, and Siberia.

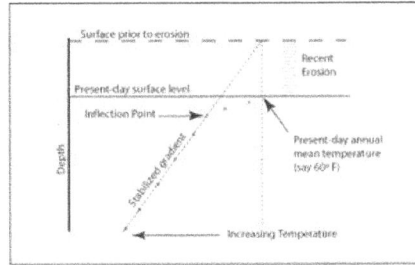

Fig 1. Borehole geothermal gradient in an area of uplift and erosion.

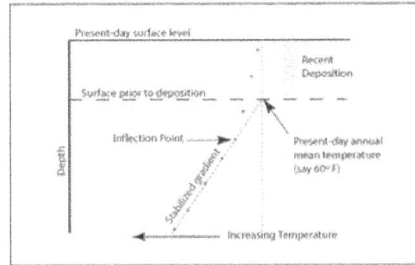

Fig 2. Borehole geothermal gradient in an area of deposition and subsidence.

In areas of Holocene uplift and erosion (Fig. 1) the initial gradient will be higher than the average until it reaches an inflection point where it reaches the stabilized heat-flow regime. If the gradient of the stabilized regime is projected above the inflection point to its intersect with present-day annual average temperature, the height of this intersect above present-day surface level gives a measure of the extent of Holocene uplift and erosion. In areas of Holocene subsidence and deposition (Fig. 2) the initial gradient will be lower than the average until it reaches an inflection point where it joins the stabilized heat-flow regime.

In deep boreholes, the temperature of the rock below the inflection point generally increases with depth at rates of the order of 20 K/km or more. Fourier's law of heat flow applied to the Earth gives $q = Mg$ where q is the heat flux at a point on the Earth's surface, M the thermal conductivity of the rocks there, and g the measured geothermal gradient. A representative value for the thermal conductivity of granitic rocks is $M = 3.0$ W/mK. Hence, using the global average geothermal conducting gradient of 0.02 K/m we get that $q = 0.06$ W/m². This estimate, corroborated by thousands of observations of heat flow in boreholes all over the world, gives a global average of 6×10^{-2} W/m². Thus, if the geothermal heat flow rising through an acre of granite terrain could be efficiently captured, it would light four 60 watt light bulbs.

A variation in surface temperature induced by climate changes and the Milankovitch cycle can penetrate below the Earth's surface and produce an oscillation in the geothermal gradient with periods varying from daily to tens of thousands of years and an amplitude which decreases with depth and having a scale depth of several kilometers. Melt water from the polar ice caps flowing along ocean bottoms tends to maintain a constant geothermal gradient throughout the Earth's surface.

If that rate of temperature change were constant, temperatures deep in the Earth would soon reach the point where all known rocks would eventually melt. We know, however, that the

Earth's mantle is solid because of the transmission of S-waves. The temperature gradient dramatically decreases with depth for two reasons. First, radioactive heat production is concentrated within the crust of the Earth, and particularly within the upper part of the crust, as concentrations of uranium, thorium, and potassium are highest there: these three elements are the main producers of radioactive heat within the Earth. Second, the mechanism of thermal transport changes from conduction, as within the rigid tectonic plates, to convection, in the portion of Earth's mantle that convects. Despite its solidity, most of the Earth's mantle behaves over long time-scales as a fluid, and heat is transported by advection, or material transport. Thus, the geothermal gradient within the bulk of Earth's mantle is of the order of 0.5 kelvin per kilometer, and is determined by the adiabatic gradient associated with mantle material (peridotite in the upper mantle).

This heating up can be both beneficial or detrimental in terms of engineering: Geothermal energy can be used as a means for generating electricity, by using the heat of the surrounding layers of rock underground to heat water and then routing the steam from this process through a turbine connected to a generator.

On the other hand, drill bits have to be cooled not only because of the friction created by the process of drilling itself but also because of the heat of the surrounding rock at great depth. Very deep mines, like some gold mines in South Africa, need the air inside to be cooled and circulated to allow miners to work at such great depth.

Geothermal Power

Geothermal power is power generated by geothermal energy. Technologies in use include dry steam power stations, flash steam power stations and binary cycle power stations. Geothermal electricity generation is currently used in 24 countries, while geothermal heating is in use in 70 countries.

As of 2015, worldwide geothermal power capacity amounts to 12.8 gigawatts (GW), of which 28 percent or 3,548 megawatts are installed in the United States. International markets grew at an average annual rate of 5 percent over the last three years and global geothermal power capacity is expected to reach 14.5–17.6 GW by 2020. Based on current geologic knowledge and technology, the Geothermal Energy Association (GEA) estimates that only 6.5 percent of total global potential has been tapped so far, while the IPCC reported geothermal power potential to be in the range of 35 GW to 2 TW. Countries generating more than 15 percent of their electricity from geothermal sources include El Salvador, Kenya, the Philippines, Iceland and Costa Rica.

Geothermal power is considered to be a sustainable, renewable source of energy because the heat extraction is small compared with the Earth's heat content. The greenhouse gas emissions of geothermal electric stations are on average 45 grams of carbon dioxide per kilowatt-hour of electricity, or less than 5 percent of that of conventional coal-fired plants.

Krafla, a geothermal power station in Iceland

Countries with installed and/or developing geothermal power projects

History and Development

In the 20th century, demand for electricity led to the consideration of geothermal power as a generating source. Prince Piero Ginori Conti tested the first geothermal power generator on 4 July 1904 in Larderello, Italy. It successfully lit four light bulbs. Later, in 1911, the world's first commercial geothermal power station was built there. Experimental generators were built in Beppu, Japan and the Geysers, California, in the 1920s, but Italy was the world's only industrial producer of geothermal electricity until 1958.

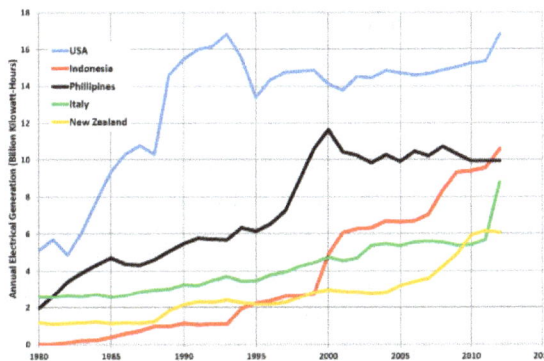

Trends in the top five geothermal electricity-generating countries, 1980–2012 (US EIA)

In 1958, New Zealand became the second major industrial producer of geothermal electricity when its Wairakei station was commissioned. Wairakei was the first station to use flash steam technology.

In 1960, Pacific Gas and Electric began operation of the first successful geothermal electric power station in the United States at The Geysers in California. The original turbine lasted for more than 30 years and produced 11 MW net power.

The binary cycle power station was first demonstrated in 1967 in Russia and later introduced to the USA in 1981, following the 1970s energy crisis and significant changes in regulatory policies. This technology allows the use of much lower temperature resources than were previously recoverable. In 2006, a binary cycle station in Chena Hot Springs, Alaska, came on-line, producing electricity from a record low fluid temperature of 57 °C (135 °F).

Geothermal electric stations have until recently been built exclusively where high temperature geothermal resources are available near the surface. The development of binary cycle power plants and improvements in drilling and extraction technology may enable enhanced geothermal systems over a much greater geographical range. Demonstration projects are operational in Landau-Pfalz, Germany, and Soultz-sous-Forêts, France, while an earlier effort in Basel, Switzerland was shut down after it triggered earthquakes. Other demonstration projects are under construction in Australia, the United Kingdom, and the United States of America.

The thermal efficiency of geothermal electric stations is low, around 7–10%, because geothermal fluids are at a low temperature compared with steam from boilers. By the laws of thermodynamics this low temperature limits the efficiency of heat engines in extracting useful energy during the generation of electricity. Exhaust heat is wasted, unless it can be used directly and locally, for example in greenhouses, timber mills, and district heating. The efficiency of the system does not affect operational costs as it would for a coal or other fossil fuel plant, but it does factor into the viability of the station. In order to produce more energy than the pumps consume, electricity generation requires high temperature geothermal fields and specialized heat cycles. Because geothermal power does not rely on variable sources of energy, unlike, for example, wind or solar, its capacity factor can be quite large – up to 96% has been demonstrated. However the global average capacity factor was 74.5% in 2008, according to the IPCC.

Resources

The earth's heat content is about 10^{31} joules. This heat naturally flows to the surface by conduction at a rate of 44.2 terawatts (TW) and is replenished by radioactive decay at a rate of 30 TW. These power rates are more than double humanity's current energy consumption from primary sources, but most of this power is too diffuse (approximately 0.1 W/m² on average) to be recoverable. The Earth's crust effectively acts as a thick insulating blanket which must be pierced by fluid conduits (of magma, water or other) to release the heat underneath.

Electricity generation requires high temperature resources that can only come from deep underground. The heat must be carried to the surface by fluid circulation, either through magma conduits, hot springs, hydrothermal circulation, oil wells, drilled water wells, or a combination of these. This circulation sometimes exists naturally where the crust is thin: magma conduits bring heat close to the surface, and hot springs bring the heat to the surface. If no hot spring is available, a well must be drilled into a hot aquifer. Away from tectonic plate boundaries the geothermal gradient is 25–30 °C per kilometre (km) of depth in most of the world, and wells would have to be several kilometres deep to permit electricity generation. The quantity and quality of recoverable resources improves with drilling depth and proximity to tectonic plate boundaries.

In ground that is hot but dry, or where water pressure is inadequate, injected fluid can stimulate production. Developers bore two holes into a candidate site, and fracture the rock between them with explosives or high pressure water. Then they pump water or liquefied carbon dioxide down one borehole, and it comes up the other borehole as a gas. This approach is called hot dry rock geothermal energy in Europe, or enhanced geothermal systems in North America. Much greater potential may be available from this approach than from conventional tapping of natural aquifers.

Estimates of the electricity generating potential of geothermal energy vary from 35 to 2000 GW depending on the scale of investments. This does not include non-electric heat recovered by co-generation, geothermal heat pumps and other direct use. A 2006 report by the Massachusetts Institute of Technology (MIT), that included the potential of enhanced geothermal systems, estimated that investing 1 billion US dollars in research and development over 15 years would allow the creation of 100 GW of electrical generating capacity by 2050 in the United States alone. The MIT report estimated that over 200 zettajoules (ZJ) would be extractable, with the potential to increase this to over 2,000 ZJ with technology improvements – sufficient to provide all the world's present energy needs for several millennia.

At present, geothermal wells are rarely more than 3 kilometres (1.9 mi) deep. Upper estimates of geothermal resources assume wells as deep as 10 kilometres (6.2 mi). Drilling near this depth is now possible in the petroleum industry, although it is an expensive process. The deepest research well in the world, the Kola superdeep borehole, is 12.3 km (7.6 mi) deep. This record has recently been imitated by commercial oil wells, such as Exxon's Z-12 well in the Chayvo field, Sakhalin. Wells drilled to depths greater than 4 kilometres (2.5 mi) generally incur drilling costs in the tens of millions of dollars. The technological challenges are to drill wide bores at low cost and to break larger volumes of rock.

Geothermal power is considered to be sustainable because the heat extraction is small compared to the Earth's heat content, but extraction must still be monitored to avoid local depletion. Although geothermal sites are capable of providing heat for many decades, individual wells may cool down or run out of water. The three oldest sites, at Larderello, Wairakei, and the Geysers have all reduced production from their peaks. It is not clear whether these stations extracted energy faster than it was replenished from greater depths, or whether the aquifers supplying them are being depleted. If production is reduced, and water is reinjected, these wells could theoretically recover their full potential. Such mitigation strategies have already been implemented at some sites. The long-term sustainability of geothermal energy has been demonstrated at the Lardarello field in Italy since 1913, at the Wairakei field in New Zealand since 1958, and at The Geysers field in California since 1960.

Power Station Types

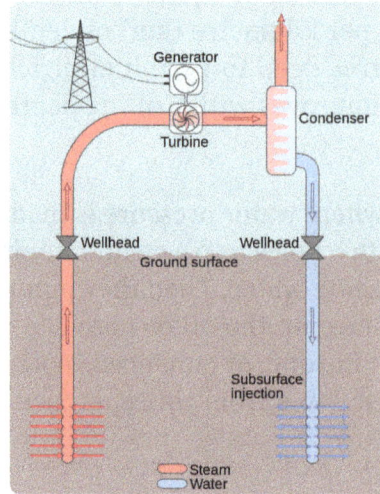

Dry steam station

Geothermal power stations are similar to other steam turbine thermal power stations – heat from a fuel source (in geothermal's case, the earth's core) is used to heat water or another working fluid. The working fluid is then used to turn a turbine of a generator, thereby producing electricity. The fluid is then cooled and returned to the heat source.

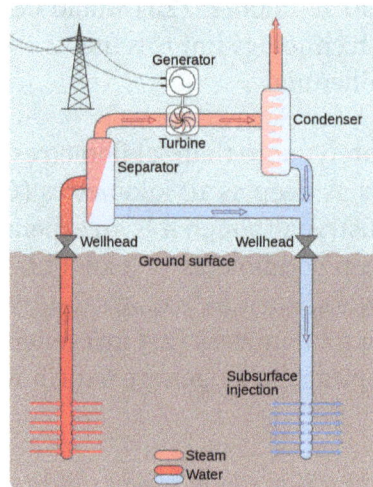

Flash steam station

Dry Steam Power Stations

Dry steam stations are the simplest and oldest design. They directly use geothermal steam of 150 °C or greater to turn turbines.

Flash Steam Power Stations

Flash steam stations pull deep, high-pressure hot water into lower-pressure tanks and use the resulting flashed steam to drive turbines. They require fluid temperatures of at least 180 °C, usually more. This is the most common type of station in operation today. Flash Steam plants use geothermal reservoirs of water with temperatures greater than 360 °F (182). The hot water flows up through wells in the ground under its own pressure. As it flows upward, the pressure decreases and some of the hot water boils into steam. The steam is then separated from the water and used to power a turbine/generator. Any leftover water and condensed steam may be injected back into the reservoir, making this a potentially sustainable resource. At The Geysers in California, twenty years of power production had depleted the groundwater and operations were substantially reduced. To restore some of the former capacity, water injection was developed.

Binary Cycle Power Stations

Binary cycle power stations are the most recent development, and can accept fluid temperatures as low as 57 °C. The moderately hot geothermal water is passed by a secondary fluid with a much lower boiling point than water. This causes the secondary fluid to flash vaporize, which then drives the turbines. This is the most common type of geothermal electricity station being constructed today. Both Organic Rankine and Kalina cycles are used. The thermal efficiency of this type station is typically about 10–13%.

Worldwide Production

The International Geothermal Association (IGA) has reported that 10,715 megawatts (MW) of geothermal power in 24 countries is online, which is expected to generate 67,246 GWh of electricity in 2010. This represents a 20% increase in geothermal power online capacity since 2005. IGA projected this would grow to 18,500 MW by 2015, due to the large number of projects that were under consideration, often in areas previously assumed to have little exploitable resource.

Larderello Geothermal Station, in Italy

In 2010, the United States led the world in geothermal electricity production with 3,086 MW of installed capacity from 77 power stations; the largest group of geothermal power plants in the world is located at The Geysers, a geothermal field in California. The Philippines follows the US as the second highest producer of geothermal power in the world, with 1,904 MW of capacity online; geothermal power makes up approximately 27% of the country's electricity generation.

Al Gore said in The Climate Project Asia Pacific Summit that Indonesia could become a super power country in electricity production from geothermal energy. India has announced a plan to develop the country's first geothermal power facility in Chhattisgarh.

Canada is the only major country on the Pacific Ring of Fire which has not yet developed geothermal power. The region of greatest potential is the Canadian Cordillera, stretching from British Columbia to the Yukon, where estimates of generating output have ranged from 1,550 MW to 5,000 MW.

Utility-grade Stations

The largest group of geothermal power plants in the world is located at The Geysers, a geothermal field in California, United States. As of 2004, five countries (El Salvador, Kenya, the Philippines, Iceland, and Costa Rica) generate more than 15% of their electricity from geothermal sources.

A geothermal power station in Negros Oriental, Philippines.

Geothermal electricity is generated in the 24 countries listed in the table below. During 2005, contracts were placed for an additional 500 MW of electrical capacity in the United States, while there were also stations under construction in 11 other countries. Enhanced geothermal systems that are several kilometres in depth are operational in France and Germany and are being developed or evaluated in at least four other countries.

Installed geothermal electric capacity					
Country	Capacity (MW) 2007	Capacity (MW) 2010	Capacity (MW) 2013	Capacity (MW) 2015	Share of national generation (%)
USA	2687	3086	3389	3450	0.3
Philippines	1969.7	1904	1894	1870	27.0
Indonesia	992	1197	1333	1340	3.7
Mexico	953	958	980	1017	3.0
New Zealand	471.6	628	895	1005	14.5
Italy	810.5	843	901	916	1.5
Iceland	421.2	575	664	665	30.0

Kenya	128.8	167	215	594	51.0
Japan	535.2	536	537	519	0.1
Turkey	38	82	163	397	0.3
Costa Rica	162.5	166	208	207	14.0
El Salvador	204.4	204	204	204	25.0
Nicaragua	87.4	88	104	159	10.0
Russia	79	82	97	82	
Papua New Guinea	56	56	56	50	
Guatemala	53	52	42	52	
Portugal	23	29	28	29	
China	27.8	24	27	27	
Germany	8.4	6.6	13	27	
France	14.7	16	15	16	
Ethiopia	7.3	7.3	8	7.3	
Austria	1.1	1.4	1	1.2	
Australia	0.2	1.1	1	1.1	
Thailand	0.3	0.3	0.3	0.3	
Total	**9,731.9**	**10,709.7**	**11,765**	**12,635.9**	–

Environmental Impact

The 120-MWe Nesjavellir power station in southwest Iceland

Fluids drawn from the deep earth carry a mixture of gases, notably carbon dioxide (CO2), hydrogen sulfide (H2S), methane (CH4), ammonia (NH3) and radon (Rn). These pollutants contribute to global warming, acid rain, radiation and noxious smells if released.

Existing geothermal electric stations, that fall within the 50th percentile of all total life cycle emissions studies reviewed by the IPCC, produce on average 45 kg of CO2 equivalent emissions per megawatt-hour of generated electricity (kg CO2eq/MW·h). For comparison, a coal-fired power plant emits 1,001 kg of CO2 per megawatt-hour when not coupled with carbon capture and storage (CCS).

Stations that experience high levels of acids and volatile chemicals are usually equipped with emission-control systems to reduce the exhaust. Geothermal stations could theoretically inject these gases back into the earth, as a form of carbon capture and storage.

In addition to dissolved gases, hot water from geothermal sources may hold in solution trace amounts of toxic chemicals, such as mercury, arsenic, boron, antimony, and salt. These chemicals come out of solution as the water cools, and can cause environmental damage if released. The modern practice of injecting geothermal fluids back into the Earth to stimulate production has the side benefit of reducing this environmental risk.

Station construction can adversely affect land stability. Subsidence has occurred in the Wairakei field in New Zealand. Enhanced geothermal systems can trigger earthquakes due to water injection. The project in Basel, Switzerland was suspended because more than 10,000 seismic events measuring up to 3.4 on the Richter Scale occurred over the first 6 days of water injection. The risk of geothermal drilling leading to uplift has been experienced in Staufen im Breisgau.

Geothermal has minimal land and freshwater requirements. Geothermal stations use 404 square meters per GW·h versus 3,632 and 1,335 square meters for coal facilities and wind farms respectively. They use 20 litres of freshwater per MW·h versus over 1000 litres per MW·h for nuclear, coal, or oil.

Geothermal power stations can also disrupt the natural cycles of geysers. For example, the Beowawe, Nevada geysers, which were uncapped geothermal wells, stopped erupting due to the development of the dual-flash station.

Economics

Geothermal power requires no fuel; it is therefore immune to fuel cost fluctuations. However, capital costs tend to be high. Drilling accounts for over half the costs, and exploration of deep resources entails significant risks. A typical well doublet in Nevada can support 4.5 megawatts (MW) of electricity generation and costs about $10 million to drill, with a 20% failure rate. In total, electrical station construction and well drilling costs about 2–5 million € per MW of electrical capacity, while the levelised energy cost is 0.04–0.10 € per kW·h. Enhanced geothermal systems tend to be on the high side of these ranges, with capital costs above $4 million per MW and levelized costs above $0.054 per kW·h in 2007.

Geothermal power is highly scalable: a small power station can supply a rural village, though initial capital costs can be high.

The most developed geothermal field is the Geysers in California. In 2008, this field supported 15 stations, all owned by Calpine, with a total generating capacity of 725 MW.

Binary Cycle

A binary cycle power plant is a type of geothermal power plant that allows cooler geothermal reservoirs to be used than is necessary for dry steam and flash steam plants. As of 2010, flash steam plants are the most common type of geothermal power generation plants in operation today, which use water at temperatures greater than 182 °C (455 K; 360 °F) that is pumped under high pressure to the generation equipment at the surface. With binary cycle geothermal power plants, pumps are used to pump hot water from a geothermal well, through a heat exchanger, and the cooled water is returned to the underground reservoir. A second "working" or "binary" fluid with a low boiling point, typically a butane or pentane hydrocarbon, is pumped at fairly high pressure (500 psi (3.4 MPa)) through the heat exchanger, where it is vaporized and then directed through a turbine. The vapor exiting the turbine is then condensed by cold air radiators or cold water and cycled back through the heat exchanger.

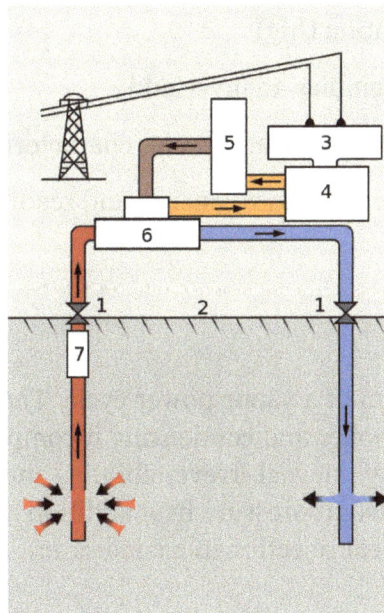

Electricity generation in a vapor-dominated hydrothermal system. Key: 1 Wellheads 2 Ground surface 3 Generator 4 Turbine 5 Condenser 6 Heat exchanger 7 Pump

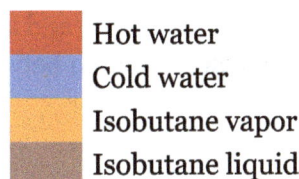

Hot water
Cold water
Isobutane vapor
Isobutane liquid

A binary vapor cycle is defined in thermodynamics as a power cycle that is a combination of two cycles, one in a high temperature region and the other in a lower temperature region.

Introduction to Binary Cycles

The use of mercury-water cycles in the United States can be dated back to the late 1920s. A small mercury-water plant which produced about 40 megawatts (MW) was in use in New Hampshire in the 1950s, with a higher thermal efficiency than most of the power plants in use during the 1950s. Unfortunately, binary vapor cycles have a high initial cost and so they are not as economically attractive.

Water is the optimal working fluid to use in vapor cycles because it is the closest to an ideal working fluid that is currently available. The binary cycle is a process designed to overcome the imperfections of water as a working fluid. The cycle uses two fluids in an attempt to approach an ideal working fluid.

Characteristics of Optimal Working Fluids

- A high critical temperature and maximum pressure

- Low triple-point temperature

- A condenser pressure that is not too low (a substance with a saturation pressure at the ambient temperature is too low)

- A high enthalpy of vaporization (hfg)

- A saturation dome that resembles an inverted U

- High thermal conductivity (good heat transfer characteristics)

- Other properties: nontoxic, inert, inexpensive, and readily available

Systems

Rankine Vapor Cycle

The Rankine cycle is the ideal form of a vapor power cycle. The ideal conditions can be reached by superheating the steam in the boiler and condensing it completely in the condenser. The ideal Rankine cycle does not involve any internal irreversibilities and consists of four processes; isentropic compression in a pump, constant pressure heat addition in a boiler, isentropic expansion in a turbine, and constant pressure heat rejection in a condenser.

Dual Pressure

This process is designed to reduce the thermodynamic losses incurred in the brine heat exchangers of the basic cycle. The losses occur through the process of transferring heat across a large temperature difference between the high temperature brine and the lower temperature of the working fluid. Losses are reduced by maintaining a closer match between the brine cooling curve and the working fluid heating curve.

Dual Fluid

"Power is extracted from a stream of hot fluid, such as geothermal water, by passing the stream in heat exchange relationship with a working fluid to vaporize the latter, expanding the vapor through a turbine, and condensing the vapor in a conventional Rankine cycle. Additional power is obtained in a second Rankine cycle by employing a portion of the hot fluid after heat exchange with the working fluid to vaporize a second working fluid having a lower boiling point and higher vapor density than the first fluid."

Power Plants

There are numerous binary cycle power stations in commercial production.

- Olkaria III, Kenya

- Mammoth Lakes, California, United States

- Steamboat Springs (Nevada), United States

- Te Huka Power Station, New Zealand

Binary cycle power plants have a thermal efficiency of 10-13%.

Geothermal Heating

Geothermal heating is the direct use of geothermal energy for heating some applications. Humans have taken advantage of geothermal heat this way since the Paleolithic era. Approximately seventy countries made direct use of a total of 270 PJ of geothermal heating in 2004. As of 2007, 28 GW of geothermal heating capacity is installed around the world, satisfying 0.07% of global primary energy consumption. Thermal efficiency is high since no energy conversion is needed, but capacity factors tend to be low (around 20%) since the heat is mostly needed in the winter.

Geothermal energy originates from the heat retained within the Earth since the original formation of the planet, from radioactive decay of minerals, and from solar energy absorbed at the surface. Most high temperature geothermal heat is harvested in regions close to tectonic plate boundaries where volcanic activity rises close to the surface of the Earth. In these areas, ground and groundwater can be found with temperatures higher than the target temperature of the application. However, even cold ground contains heat, below 6 metres (20 ft) the undisturbed ground temperature is consistently at the Mean Annual Air Temperature and it may be extracted with a heat pump.

Applications

Top countries using the most geothermal heating in 2005			
Country	Production PJ/yr	Capacity GW	Capacity Factor	Dominant applications
China	45.38	3.69	39%	bathing
Sweden	43.2	4.2	33%	heat pumps
USA	31.24	7.82	13%	heat pumps
Turkey	24.84	1.5	53%	district heating

Iceland	24.5	1.84	42%	district heating
Japan	10.3	0.82	40%	bathing (onsens)
Hungary	7.94	0.69	36%	spas/greenhouses
Italy	7.55	0.61	39%	spas/space heating
New Zealand	7.09	0.31	73%	industrial uses
63 others	71	6.8		
Total	**273**	**28**	**31%**	**space heating**

There are a wide variety of applications for cheap geothermal heat. In 2004 more than half of direct geothermal heat was used for space heating, and a third was used for spas. The remainder was used for a variety of industrial processes, desalination, domestic hot water, and agricultural applications. The cities of Reykjavík and Akureyri pipe hot water from geothermal plants under roads and pavements to melt snow. Geothermal desalination has been demonstrated.

Geothermal systems tend to benefit from economies of scale, so space heating power is often distributed to multiple buildings, sometimes whole communities. This technique, long practiced throughout the world in locations such as Reykjavík, Iceland, Boise, Idaho, and Klamath Falls, Oregon is known as district heating.

Extraction

Some parts of the world, including substantial portions of the western USA, are underlain by relatively shallow geothermal resources. Similar conditions exist in Iceland, parts of Japan, and other geothermal hot spots around the world. In these areas, water or steam may be captured from natural hot springs and piped directly into radiators or heat exchangers. Alternatively, the heat may come from waste heat supplied by co-generation from a geothermal electrical plant or from deep wells into hot aquifers. Direct geothermal heating is far more efficient than geothermal electricity generation and has less demanding temperature requirements, so it is viable over a large geographical range. If the shallow ground is hot but dry, air or water may be circulated through earth tubes or downhole heat exchangers which act as heat exchangers with the ground.

Steam under pressure from deep geothermal resources is also used to generate electricity from geothermal power. The Iceland Deep Drilling Project struck a pocket of magma at 2,100m. A cemented steelcase was constructed in the hole with a perforation at the bottom close to the magma. The high temperatures and pressure of the magma steam were used to generate 36MW of electricity, making IDDP-1 the world's first magma-enhanced geothermal system.

In areas where the shallow ground is too cold to provide comfort directly, it is still warmer than the winter air. The thermal inertia of the shallow ground retains solar energy accumulated in the summertime, and seasonal variations in ground temperature disappear completely below 10m of depth. That heat can be extracted with a geothermal heat pump more efficiently than it can be generated by conventional furnaces. Geothermal heat pumps are economically viable essentially anywhere in the world.

Ground-source Heat Pumps

In regions without any high temperature geothermal resources, a ground-source heat pump (GSHP) can provide space heating and space cooling. Like a refrigerator or air conditioner, these systems use a heat pump to force the transfer of heat from the ground to the building. Heat can be extracted from any source, no matter how cold, but a warmer source allows higher efficiency. A ground-source heat pump uses the shallow ground or ground water (typically starting at 10–12 °C or 50–54 °F) as a source of heat, thus taking advantage of its seasonally moderate temperatures. In contrast, an air-source heat pump draws heat from the air (colder outside air) and thus requires more energy.

Ground source heat pumps (GSHPs) are *not* geothermal, i.e. there is no geyser providing heat to be captured. A GSHP merely accesses stored solar heat energy in the soil or rock. GSHPs circulate a carrier fluid (usually a mixture of water and small amounts of antifreeze) through closed pipe loops buried in the ground. Single-home systems can be "vertical loop field" systems with bore holes 50–400 feet (15–120 m) deep or, if adequate land is available for extensive trenches, a "horizontal loop field" is installed approximately six feet subsurface. As the fluid circulates underground it absorbs heat from the ground and, on its return, the warmed fluid passes through the heat pump which uses electricity to extract heat from the fluid. The re-chilled fluid is sent back into the ground thus continuing the cycle. The heat extracted and that generated by the heat pump appliance as a byproduct is used to heat the house. The addition of the ground heating loop in the energy equation means that significantly more heat can be transferred to a building than if electricity alone had been used directly for heating.

Switching the direction of heat flow, the same system can be used to circulate the cooled water through the house for cooling in the summer months. The heat is exhausted to the relatively cooler ground (or groundwater) rather than delivering it to the hot outside air as an air conditioner does. As a result, the heat is pumped across a larger temperature difference and this leads to higher efficiency and lower energy use.

This technology makes ground source heating economically viable in any geographical location. In 2004, an estimated million ground-source heat pumps with a total capacity of 15 GW extracted 88 PJ of heat energy for space heating. Global ground-source heat pump capacity is growing by 10% annually.

History

Hot springs have been used for bathing at least since Paleolithic times. The oldest known spa is a stone pool on China's Mount Li built in the Qin dynasty in the 3rd century BC, at the same site where the Huaqing Chi palace was later built. Geothermal energy supplied channeled district heating for baths and houses in Pompeii around 0 AD. In the first century AD, Romans conquered Aquae Sulis in England and used the hot springs there to feed public baths and underfloor heating. The admission fees for these baths probably represents the first commercial use of geothermal power. A 1,000-year-old hot tub has been located in Iceland, where it was built by one of the island's original settlers. The world's oldest working geothermal district heating system in Chaudes-Aigues, France, has been operating since the 14th century. The earliest industrial exploitation began in 1827 with the use of geyser steam to extract boric acid from volcanic mud in Larderello, Italy.

In 1892, America's first district heating system in Boise, Idaho was powered directly by geothermal energy, and was soon copied in Klamath Falls, Oregon in 1900. A deep geothermal well was used to heat greenhouses in Boise in 1926, and geysers were used to heat greenhouses in Iceland and Tuscany at about the same time. Charlie Lieb developed the first downhole heat exchanger in 1930 to heat his house. Steam and hot water from the geysers began to be used to heat homes in Iceland in 1943.

By this time, Lord Kelvin had already invented the heat pump in 1852, and Heinrich Zoelly had patented the idea of using it to draw heat from the ground in 1912. But it was not until the late 1940s that the geothermal heat pump was successfully implemented. The earliest one was probably Robert C. Webber's home-made 2.2 kW direct-exchange system, but sources disagree as to the exact timeline of his invention. J. Donald Kroeker designed the first commercial geothermal heat pump to heat the Commonwealth Building (Portland, Oregon) and demonstrated it in 1946. Professor Carl Nielsen of Ohio State University built the first residential open loop version in his home in 1948. The technology became popular in Sweden as a result of the 1973 oil crisis, and has been growing slowly in worldwide acceptance since then. The 1979 development of polybutylene pipe greatly augmented the heat pump's economic viability. As of 2004, there are over a million geothermal heat pumps installed worldwide providing 12 GW of thermal capacity. Each year, about 80,000 units are installed in the USA and 27,000 in Sweden.

Economics

Geothermal energy is a type of renewable energy that encourages conservation of natural resources. According to the U.S. Environmental Protection Agency, geo-exchange systems save homeowners 30–70 percent in heating costs, and 20–50 percent in cooling costs, compared to conventional systems. Geo-exchange systems also save money because they require much less maintenance. In addition to being highly reliable they are built to last for decades.

Some utilities, such as Kansas City Power and Light, offer special, lower winter rates for geothermal customers, offering even more savings.

Geothermal Drilling Risks

In geothermal heating projects the underground is penetrated by trenches or drillholes. As with all underground work, projects may cause problems if the geology of the area is poorly understood.

Cracks at the historic Town Hall of Staufen im Breisgau presumed due to damage from geothermal drilling

In the spring of 2007 an exploratory geothermal drilling operation was conducted to provide geothermal heat to the town hall of Staufen im Breisgau. After initially sinking a few millimeters, a process called subsidence, the city center has started to rise gradually causing considerable damage to buildings in the city center, affecting numerous historic houses including the town hall. It is hypothesized that the drilling perforated an anhydrite layer bringing high-pressure groundwater to come into contact with the anhydrite, which then began to expand. Currently no end to the rising process is in sight. Data from the TerraSAR-X radar satellite before and after the changes confirmed the localised nature of the situation.

A geochemical process called anhydrite swelling has been confirmed as the cause of these uplifts. This is a transformation of the mineral anhydrite (anhydrous calcium sulphate) into gypsum (hydrous calcium sulphate). A pre-condition for this transformation is that the anhydrite is in contact with water, which is then stored in its crystalline structure.

Geothermal Heat Pump

A geothermal heat pump or ground source heat pump (GSHP) is a central heating and/or cooling system that transfers heat to or from the ground.

It uses the earth as a heat source (in the winter) or a heat sink (in the summer). This design takes advantage of the moderate temperatures in the ground to boost efficiency and reduce the operational costs of heating and cooling systems, and may be combined with solar heating to form a geosolar system with even greater efficiency. They are also known by other names, including geoexchange, earth-coupled, earth energy systems. The engineering and scientific communities prefer the terms "geoexchange" or "ground source heat pumps" to avoid confusion with traditional geothermal power, which uses a high temperature heat source to generate electricity. Ground source heat pumps harvest heat absorbed at the Earth's surface from solar energy. The temperature in the ground below 6 metres (20 ft) is roughly equal to the mean annual air temperature at that latitude at the surface.

Depending on latitude, the temperature beneath the upper 6 metres (20 ft) of Earth's surface maintains a nearly constant temperature between 10 and 16 °C (50 and 60 °F), if the temperature is undisturbed by the presence of a heat pump. Like a refrigerator or air conditioner, these systems use a heat pump to force the transfer of heat from the ground. Heat pumps can transfer heat from a cool space to a warm space, against the natural direction of flow, or they can enhance the natural flow of heat from a warm area to a cool one. The core of the heat pump is a loop of refrigerant pumped through a vapor-compression refrigeration cycle that moves heat. Air-source heat pumps are typically more efficient at heating than pure electric heaters, even when extracting heat from cold winter air, although efficiencies begin dropping significantly as outside air temperatures drop below 5 °C (41 °F). A ground source heat pump exchanges heat with the ground. This is much more energy-efficient because underground temperatures are more stable than air temperatures through the year. Seasonal variations drop off with depth and disappear below 7 metres (23 ft) to 12 metres (39 ft) due to thermal inertia. Like a cave, the shallow ground temperature is warmer than the air above during the winter and cooler than the air in the summer. A ground source heat pump extracts ground heat in the winter (for heating) and transfers heat back into the ground in the summer (for cooling). Some systems are designed to operate in one mode only, heating or cooling, depending on climate.

Geothermal pump systems reach fairly high coefficient of performance (CoP), 3 to 6, on the coldest of winter nights, compared to 1.75–2.5 for air-source heat pumps on cool days. Ground source heat pumps (GSHPs) are among the most energy efficient technologies for providing HVAC and water heating.

Setup costs are higher than for conventional systems, but the difference is usually returned in energy savings in 3 to 10 years, and even shorter lengths of time with federal, state and utility tax credits and incentives. Geothermal heat pump systems are reasonably warranted by manufacturers, and their working life is estimated at 25 years for inside components and 50+ years for the ground loop. As of 2004, there are over a million units installed worldwide providing 12 GW of thermal capacity, with an annual growth rate of 10%.

Differing Terms and Definitions

Ground source heating and cooling

Some confusion exists with regard to the terminology of heat pumps and the use of the term "*geothermal*". "*Geothermal*" derives from the Greek and means "*Earth heat*" - which geologists and many laymen understand as describing hot rocks, volcanic activity or heat derived from deep within the earth. Though some confusion arises when the term "*geothermal*" is also used to apply to temperatures within the first 100 metres of the surface, this is "*Earth heat*" all the same, though it is largely influenced by stored energy from the sun.

History

The heat pump was described by Lord Kelvin in 1853 and developed by Peter Ritter von Rittinger in 1855. After experimenting with a freezer, Robert C. Webber built the first direct exchange ground-source heat pump in the late 1940s. The first successful commercial project was installed in the Commonwealth Building (Portland, Oregon) in 1948, and has been designated a National Historic Mechanical Engineering Landmark by ASME. The technology became popular in Sweden in the 1970s, and has been growing slowly in worldwide acceptance since then. Open loop systems dominated the market until the development of polybutylene pipe in 1979 made closed loop systems economically viable. As of 2004, there are over a million units installed worldwide providing 12 GW of thermal capacity. Each year, about 80,000 units are installed in the US (geothermal energy is used in all 50 U.S. states today, with great potential for near-term market growth and savings) and 27,000 in Sweden. In Finland, a geothermal heat pump was the most common heating system choice for new detached houses between 2006 and 2011 with market share exceeding 40%.

Ground Heat Exchanger

Heat pumps provide winter heating by extracting heat from a source and transferring it into a building. Heat can be extracted from any source, no matter how cold, but a warmer source allows higher efficiency. A ground source heat pump uses the top layer of the earth's crust as a source of heat, thus taking advantage of its seasonally moderated temperature.

Loop field for a 12-ton system (unusually large for most residential applications)

In the summer, the process can be reversed so the heat pump extracts heat from the building and transfers it to the ground. Transferring heat to a cooler space takes less energy, so the cooling efficiency of the heat pump gains benefits from the lower ground temperature.

Ground source heat pumps employ a heat exchanger in contact with the ground or groundwater to extract or dissipate heat. This component accounts for anywhere from a fifth to half of the total system cost, and would be the most cumbersome part to repair or replace. Correctly sizing this component is necessary to assure long-term performance: the energy efficiency of the system improves with roughly 4% for every degree Celsius that is won through correct sizing, and the underground temperature balance must be maintained through proper design of the whole system. Incorrect design can result in the system freezing after a number of years or very inefficient system performance; thus accurate system design is critical to a successful system

Shallow 3–8-foot (0.91–2.44 m) horizontal heat exchangers experience seasonal temperature cycles due to solar gains and transmission losses to ambient air at ground level. These temperature cycles lag behind the seasons because of thermal inertia, so the heat exchanger will harvest heat deposited by the sun several months earlier, while being weighed down in late winter and spring, due to accumulated winter cold. Deep vertical systems 100–500 feet (30–152 m) deep rely on migration of heat from surrounding geology, unless they are recharged annually by solar recharge of the ground or exhaust heat from air conditioning systems.

Several major design options are available for these, which are classified by fluid and layout. Direct exchange systems circulate refrigerant underground, closed loop systems use a mixture of anti-freeze and water, and open loop systems use natural groundwater.

Direct Exchange

The direct exchange geothermal heat pump is the oldest type of geothermal heat pump technology. The ground-coupling is achieved through a single loop, circulating refrigerant, in direct thermal contact with the ground (as opposed to a combination of a refrigerant loop and a water loop). The refrigerant leaves the heat pump cabinet, circulates through a loop of copper tube buried underground, and exchanges heat with the ground before returning to the pump. The name "direct exchange" refers to heat transfer between the refrigerant loop and the ground without the use of an intermediate fluid. There is no direct interaction between the fluid and the earth; only heat transfer through the pipe wall. Direct exchange heat pumps, which are now rarely used, are not to be confused with "water-source heat pumps" or "water loop heat pumps" since there is no water in the ground loop. ASHRAE defines the term ground-coupled heat pump to encompass closed loop and direct exchange systems, while excluding open loops.

Direct exchange systems are more efficient and have potentially lower installation costs than closed loop water systems. Copper's high thermal conductivity contributes to the higher efficiency of the system, but heat flow is predominantly limited by the thermal conductivity of the ground, not the pipe. The main reasons for the higher efficiency are the elimination of the water pump (which uses electricity), the elimination of the water-to-refrigerant heat exchanger (which is a source of heat losses), and most importantly, the latent heat phase change of the refrigerant in the ground itself.

While they require more refrigerant and their tubing is more expensive per foot, a direct exchange earth loop is shorter than a closed water loop for a given capacity. A direct exchange system requires only 15 to 30% of the length of tubing and half the diameter of drilled holes, and the drilling or excavation costs are therefore lower. Refrigerant loops are less tolerant of leaks than water loops because gas can leak out through smaller imperfections. This dictates the use of brazed copper tubing, even though the pressures are similar to water loops. The copper loop must be protected from corrosion in acidic soil through the use of a sacrificial anode or other cathodic protection.

The U.S. Environmental Protection Agency conducted field monitoring of a direct geoexchange heat pump water heating system in a commercial application. The EPA reported that the system saved 75% of the electrical energy that would have been required by an electrical resistance water heating unit. According to the EPA, if the system is operated to capacity, it can avoid the emission of up to 7,100 pounds of CO_2 and 15 pounds of NO_x each year per ton of compressor capacity (or 42,600 lbs. of CO_2 and 90 lbs. of NO_x for a typical 6 ton system).

In Northern climates, although the earth temperature is cooler, so is the incoming water temperature, which enables the high efficiency systems to replace more energy than would otherwise be required of electric or fossil fuel fired systems. Any temperature above -40 °F is sufficient to evaporate the refrigerant, and the direct exchange system can harvest energy through ice.

In extremely hot climates with dry soil, the addition of an auxiliary cooling module as a second condenser in line between the compressor and the earth loops increases efficiency and can further reduce the amount of earth loop to be installed.

Closed Loop

Most installed systems have two loops on the ground side: the primary refrigerant loop is contained in the appliance cabinet where it exchanges heat with a secondary water loop that is buried underground. The secondary loop is typically made of high-density polyethylene pipe and contains a mixture of water and anti-freeze (propylene glycol, denatured alcohol or methanol). Monopropylene glycol has the least damaging potential when it might leak into the ground, and is therefore the only allowed anti-freeze in ground sources in an increasing number of European countries. After leaving the internal heat exchanger, the water flows through the secondary loop outside the building to exchange heat with the ground before returning. The secondary loop is placed below the frost line where the temperature is more stable, or preferably submerged in a body of water if available. Systems in wet ground or in water are generally more efficient than drier ground loops since water conducts and stores heat better than solids in sand or soil. If the ground is naturally dry, soaker hoses may be buried with the ground loop to keep it wet.

An installed liquid pump pack

Closed loop systems need a heat exchanger between the refrigerant loop and the water loop, and pumps in both loops. Some manufacturers have a separate ground loop fluid pump pack, while some integrate the pumping and valving within the heat pump. Expansion tanks and pressure relief valves may be installed on the heated fluid side. Closed loop systems have lower efficiency than direct exchange systems, so they require longer and larger pipe to be placed in the ground, increasing excavation costs.

Closed loop tubing can be installed horizontally as a loop field in trenches or vertically as a series of long U-shapes in wells. The size of the loop field depends on the soil type and moisture content, the average ground temperature and the heat loss and or gain characteristics of the building being conditioned. A rough approximation of the initial soil temperature is the average daily temperature for the region.

Vertical

A vertical closed loop field is composed of pipes that run vertically in the ground. A hole is bored in the ground, typically 50 to 400 feet (15–122 m) deep. Pipe pairs in the hole are joined with a

U-shaped cross connector at the bottom of the hole. The borehole is commonly filled with a bentonite grout surrounding the pipe to provide a thermal connection to the surrounding soil or rock to improve the heat transfer. Thermally enhanced grouts are available to improve this heat transfer. Grout also protects the ground water from contamination, and prevents artesian wells from flooding the property. Vertical loop fields are typically used when there is a limited area of land available. Bore holes are spaced at least 5–6 m apart and the depth depends on ground and building characteristics. For illustration, a detached house needing 10 kW (3 ton) of heating capacity might need three boreholes 80 to 110 m (260 to 360 ft) deep. (A ton of heat is 12,000 British thermal units per hour (BTU/h) or 3.5 kilowatts.) During the cooling season, the local temperature rise in the bore field is influenced most by the moisture travel in the soil. Reliable heat transfer models have been developed through sample bore holes as well as other tests.

Drilling of a borehole for residential heating

Horizontal

A horizontal closed loop field is composed of pipes that run horizontally in the ground. A long horizontal trench, deeper than the frost line, is dug and U-shaped or slinky coils are placed horizontally inside the same trench. Excavation for shallow horizontal loop fields is about half the cost of vertical drilling, so this is the most common layout used wherever there is adequate land available. For illustration, a detached house needing 10 kW (3 ton) of heating capacity might need three loops 120 to 180 m (390 to 590 ft) long of NPS 3/4 (DN 20) or NPS 1.25 (DN 32) polyethylene tubing at a depth of 1 to 2 m (3.3 to 6.6 ft).

The depth at which the loops are placed significantly influences the energy consumption of the heat pump in two opposite ways: shallow loops tend to indirectly absorb more heat from the sun, which is helpful, especially when the ground is still cold after a long winter. On the other hand, shallow loops are also cooled down much more readily by weather changes, especially during long cold winters, when heating demand peaks. Often, the second effect is much greater than the first one, leading to higher costs of operation for the more shallow ground loops. This problem can be reduced by increasing both the depth and the length of piping, thereby significantly increasing costs of installation. However, such expenses might be deemed feasible, as they may result in lower operating costs. Recent studies show that utilization of a non-homogeneous soil profile with a layer of low conductive material above the ground pipes can help mitigate the adverse effects of shallow pipe burial depth. The intermediate blanket with lower conductivity than the surrounding soil profile demonstrated the potential to increase the energy extraction rates from the ground to as high as 17% for a cold climate and about 5-6% for a relatively moderate climate.

A three-ton slinky loop prior to being covered with soil. The three slinky loops are running out horizontally with three straight lines returning the end of the slinky coil to the heat pump.

A slinky (also called coiled) closed loop field is a type of horizontal closed loop where the pipes overlay each other (not a recommended method). The easiest way of picturing a slinky field is to imagine holding a slinky on the top and bottom with your hands and then moving your hands in opposite directions. A slinky loop field is used if there is not adequate room for a true horizontal system, but it still allows for an easy installation. Rather than using straight pipe, slinky coils use overlapped loops of piping laid out horizontally along the bottom of a wide trench. Depending on soil, climate and the heat pump's run fraction, slinky coil trenches can be up to two thirds shorter than traditional horizontal loop trenches. Slinky coil ground loops are essentially a more economical and space efficient version of a horizontal ground loop.

Radial or Directional Drilling

As an alternative to trenching, loops may be laid by mini horizontal directional drilling (mini-HDD). This technique can lay piping under yards, driveways, gardens or other structures without disturbing them, with a cost between those of trenching and vertical drilling. This system also differs from horizontal & vertical drilling as the loops are installed from one central chamber, further reducing the ground space needed. Radial drilling is often installed retroactively (after the property has been built) due to the small nature of the equipment used and the ability to bore beneath existing constructions.

Pond

A closed pond loop is not common because it depends on proximity to a body of water, where an open loop system is usually preferable. A pond loop may be advantageous where poor water quality precludes an open loop, or where the system heat load is small. A pond loop consists of coils of pipe similar to a slinky loop attached to a frame and located at the bottom of an appropriately sized pond or water source.

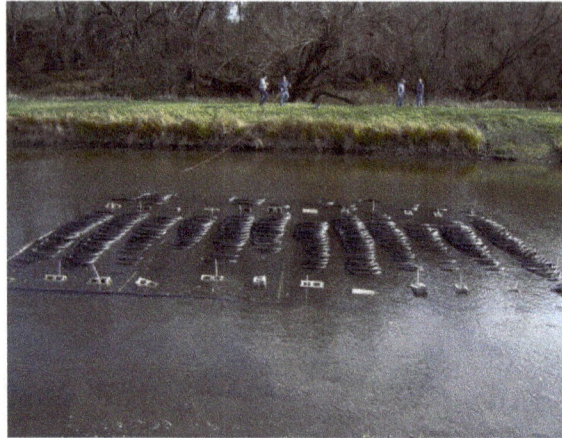

12-ton pond loop system being sunk to the bottom of a pond

Open loop

In an open loop system (also called a groundwater heat pump), the secondary loop pumps natural water from a well or body of water into a heat exchanger inside the heat pump. ASHRAE calls open loop systems *groundwater heat pumps* or *surface water heat pumps*, depending on the source. Heat is either extracted or added by the primary refrigerant loop, and the water is returned to a separate injection well, irrigation trench, tile field or body of water. The supply and return lines must be placed far enough apart to ensure thermal recharge of the source. Since the water chemistry is not controlled, the appliance may need to be protected from corrosion by using different metals in the heat exchanger and pump. Limescale may foul the system over time and require periodic acid cleaning. This is much more of a problem with cooling systems than heating systems. Also, as fouling decreases the flow of natural water, it becomes difficult for the heat pump to exchange building heat with the groundwater. If the water contains high levels of salt, minerals, iron bacteria or hydrogen sulfide, a closed loop system is usually preferable.

Deep lake water cooling uses a similar process with an open loop for air conditioning and cooling. Open loop systems using ground water are usually more efficient than closed systems because they are better coupled with ground temperatures. Closed loop systems, in comparison, have to transfer heat across extra layers of pipe wall and dirt.

A growing number of jurisdictions have outlawed open-loop systems that drain to the surface because these may drain aquifers or contaminate wells. This forces the use of more environmentally sound injection wells or a closed loop system.

Standing Column Well

A standing column well system is a specialized type of open loop system. Water is drawn from the bottom of a deep rock well, passed through a heat pump, and returned to the top of the well, where traveling downwards it exchanges heat with the surrounding bedrock. The choice of a standing column well system is often dictated where there is near-surface bedrock and limited surface area is available. A standing column is typically not suitable in locations where the geology is mostly clay, silt, or sand. If bedrock is deeper than 200 feet (61 m) from the surface, the cost of casing to seal off the overburden may become prohibitive.

A multiple standing column well system can support a large structure in an urban or rural application. The standing column well method is also popular in residential and small commercial applications. There are many successful applications of varying sizes and well quantities in the many boroughs of New York City, and is also the most common application in the New England states. This type of ground source system has some heat storage benefits, where heat is rejected from the building and the temperature of the well is raised, within reason, during the summer cooling months which can then be harvested for heating in the winter months, thereby increasing the efficiency of the heat pump system. As with closed loop systems, sizing of the standing column system is critical in reference to the heat loss and gain of the existing building. As the heat exchange is actually with the bedrock, using water as the transfer medium, a large amount of production capacity (water flow from the well) is not required for a standing column system to work. However, if there is adequate water production, then the thermal capacity of the well system can be enhanced by discharging a small percentage of system flow during the peak Summer and Winter months.

Since this is essentially a water pumping system, standing column well design requires critical considerations to obtain peak operating efficiency. Should a standing column well design be misapplied, leaving out critical shut-off valves for example, the result could be an extreme loss in efficiency and thereby cause operational cost to be higher than anticipated.

Building Distribution

The heat pump is the central unit that becomes the heating and cooling plant for the building. Some models may cover space heating, space cooling, (space heating via conditioned air, hydronic systems and / or radiant heating systems), domestic or pool water preheat (via the desuperheater function), demand hot water, and driveway ice melting all within one appliance with a variety of options with respect to controls, staging and zone control. The heat may be carried to its end use by circulating water or forced air. Almost all types of heat pumps are produced for commercial and residential applications.

Liquid-to-air heat pump

Liquid-to-air heat pumps (also called *water-to-air*) output forced air, and are most commonly used to replace legacy forced air furnaces and central air conditioning systems. There are variations that allow for split systems, high-velocity systems, and ductless systems. Heat pumps cannot achieve as high a fluid temperature as a conventional furnace, so they require a higher volume flow rate of air to compensate. When retrofitting a residence, the existing duct work may have to be enlarged to reduce the noise from the higher air flow.

Liquid-to-water heat pumps (also called *water-to-water*) are hydronic systems that use water to carry heating or cooling through the building. Systems such as radiant underfloor heating, baseboard radiators, conventional cast iron radiators would use a liquid-to-water heat pump. These heat pumps are preferred for pool heating or domestic hot water pre-heat. Heat pumps can only heat water to about 50 °C (122 °F) efficiently, whereas a boiler normally reaches 65–95 °C (149–203 °F). Legacy radiators designed for these higher temperatures may have to be doubled in numbers when retrofitting a home. A hot water tank will still be needed to raise water temperatures above the heat pump's maximum, but pre-heating will save 25–50% of hot water costs.

Liquid-to-water heat pump

Ground source heat pumps are especially well matched to underfloor heating and baseboard radiator systems which only require warm temperatures 40 °C (104 °F) to work well. Thus they are ideal for open plan offices. Using large surfaces such as floors, as opposed to radiators, distributes the heat more uniformly and allows for a lower water temperature. Wood or carpet floor coverings dampen this effect because the thermal transfer efficiency of these materials is lower than that of masonry floors (tile, concrete). Underfloor piping, ceiling or wall radiators can also be used for cooling in dry climates, although the temperature of the circulating water must be above the dew point to ensure that atmospheric humidity does not condense on the radiator.

Combination heat pumps are available that can produce forced air and circulating water simultaneously and individually. These systems are largely being used for houses that have a combination of air and liquid conditioning needs, for example central air conditioning and pool heating.

Seasonal Thermal Storage

The efficiency of ground source heat pumps can be greatly improved by using seasonal thermal energy storage and interseasonal heat transfer. Heat captured and stored in thermal banks in the summer can be retrieved efficiently in the winter. Heat storage efficiency increases with scale, so this advantage is most significant in commercial or district heating systems.

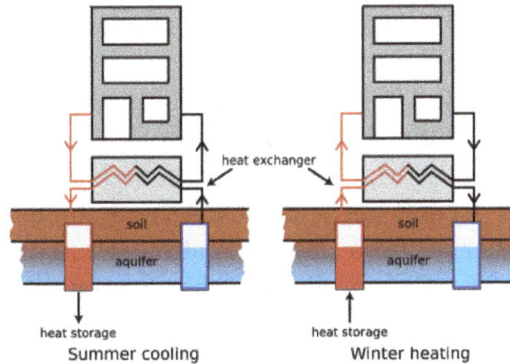

A heat pump in combination with heat and cold storage

Geosolar combisystems have been used to heat and cool a greenhouse using an aquifer for thermal storage. In summer, the greenhouse is cooled with cold ground water. This heats the water in the aquifer which can become a warm source for heating in winter. The combination of cold and heat storage with heat pumps can be combined with water/humidity regulation. These principles are used to provide renewable heat and renewable cooling to all kinds of buildings.

Also the efficiency of existing small heat pump installations can be improved by adding large, cheap, water filled solar collectors. These may be integrated into a to-be-overhauled parking lot, or in walls or roof constructions by installing one inch PE pipes into the outer layer.

Thermal Efficiency

The net thermal efficiency of a heat pump should take into account the efficiency of electricity generation and transmission, typically about 30%. Since a heat pump moves three to five times more heat energy than the electric energy it consumes, the total energy output is much greater than the electrical input. This results in net thermal efficiencies greater than 300% as compared to radiant electric heat being 100% efficient. Traditional combustion furnaces and electric heaters can never exceed 100% efficiency.

Geothermal heat pumps can reduce energy consumption— and corresponding air pollution emissions—up to 44% compared to air source heat pumps and up to 72% compared to electric resistance heating with standard air-conditioning equipment.

The dependence of net thermal efficiency on the electricity infrastructure tends to be an unnecessary complication for consumers and is not applicable to hydroelectric power, so performance of heat pumps is usually expressed as the ratio of heating output or heat removal to electricity input. Cooling performance is typically expressed in units of BTU/hr/watt as the energy efficiency ratio (EER), while heating performance is typically reduced to dimensionless units as the coefficient of performance (COP). The conversion factor is 3.41 BTU/hr/watt. Performance is influenced by all components of the installed system, including the soil conditions, the ground-coupled heat exchanger, the heat pump appliance, and the building distribution, but is largely determined by the "lift" between the input temperature and the output temperature.

For the sake of comparing heat pump appliances to each other, independently from other system components, a few standard test conditions have been established by the American Refrigerant Institute (ARI) and more recently by the International Organization for Standardization. Standard ARI 330 ratings were intended for closed loop ground-source heat pumps, and assume secondary loop water temperatures of 77 °F (25 °C) for air conditioning and 32 °F (0 °C) for heating. These temperatures are typical of installations in the northern US. Standard ARI 325 ratings were intended for open loop ground-source heat pumps, and include two sets of ratings for groundwater temperatures of 50 °F (10 °C) and 70 °F (21 °C). ARI 325 budgets more electricity for water pumping than ARI 330. Neither of these standards attempt to account for seasonal variations. Standard ARI 870 ratings are intended for direct exchange ground-source heat pumps. ASHRAE transitioned to ISO 13256-1 in 2001, which replaces ARI 320, 325 and 330. The new ISO standard produces slightly higher ratings because it no longer budgets any electricity for water pumps.

Efficient compressors, variable speed compressors and larger heat exchangers all contribute to heat pump efficiency. Residential ground source heat pumps on the market today have standard COPs ranging from 2.4 to 5.0 and EERs ranging from 10.6 to 30. To qualify for an Energy Star label, heat pumps must meet certain minimum COP and EER ratings which depend on the ground heat exchanger type. For closed loop systems, the ISO 13256-1 heating COP must be 3.3 or greater and the cooling EER must be 14.1 or greater.

Actual installation conditions may produce better or worse efficiency than the standard test conditions. COP improves with a lower temperature difference between the input and output of the heat pump, so the stability of ground temperatures is important. If the loop field or water pump is undersized, the addition or removal of heat may push the ground temperature beyond standard test conditions, and performance will be degraded. Similarly, an undersized blower may allow the plenum coil to overheat and degrade performance.

Soil without artificial heat addition or subtraction and at depths of several metres or more remains at a relatively constant temperature year round. This temperature equates roughly to the average annual air-temperature of the chosen location, usually 7–12 °C (45–54 °F) at a depth of 6 metres (20 ft) in the northern US. Because this temperature remains more constant than the air temperature throughout the seasons, geothermal heat pumps perform with far greater efficiency during extreme air temperatures than air conditioners and air-source heat pumps.

Standards ARI 210 and 240 define Seasonal Energy Efficiency Ratio (SEER) and Heating Seasonal Performance Factors (HSPF) to account for the impact of seasonal variations on air source heat pumps. These numbers are normally not applicable and should not be compared to ground source heat pump ratings. However, Natural Resources Canada has adapted this approach to calculate typical seasonally adjusted HSPFs for ground-source heat pumps in Canada. The NRC HSPFs ranged from 8.7 to 12.8 BTU/hr/watt (2.6 to 3.8 in nondimensional factors, or 255% to 375% seasonal average electricity utilization efficiency) for the most populated regions of Canada. When combined with the thermal efficiency of electricity, this corresponds to net average thermal efficiencies of 100% to 150%.

Environmental Impact

The US Environmental Protection Agency (EPA) has called ground source heat pumps the most energy-efficient, environmentally clean, and cost-effective space conditioning systems available. Heat pumps offer significant emission reductions potential, particularly where they are used for both heating and cooling and where the electricity is produced from renewable resources.

Ground-source heat pumps have unsurpassed thermal efficiencies and produce zero emissions locally, but their electricity supply includes components with high greenhouse gas emissions, unless the owner has opted for a 100% renewable energy supply. Their environmental impact therefore depends on the characteristics of the electricity supply and the available alternatives.

Country	Electricity CO_2 Emissions Intensity	GHG savings relative to		
		natural gas	heating oil	electric heating
Canada	223 ton/GWh	2.7 ton/yr	5.3 ton/yr	3.4 ton/yr
Russia	351 ton/GWh	1.8 ton/yr	4.4 ton/yr	5.4 ton/yr
US	676 ton/GWh	-0.5 ton/yr	2.2 ton/yr	10.3 ton/yr
China	839 ton/GWh	-1.6 ton/yr	1.0 ton/yr	12.8 ton/yr

Annual greenhouse gas (GHG) savings from using a ground source heat pump instead of a high-efficiency furnace in a detached residence (assuming no specific supply of renewable energy)

The GHG emissions savings from a heat pump over a conventional furnace can be calculated based on the following formula:

- HL = seasonal heat load ≈ 80 GJ/yr for a modern detached house in the northern US

- FI = emissions intensity of fuel = 50 kg(CO2)/GJ for natural gas, 73 for heating oil, 0 for 100% renewable energy such as wind, hydro, photovoltaic or solar thermal

- AFUE = furnace efficiency ≈ 95% for a modern condensing furnace

- COP = heat pump coefficient of performance ≈ 3.2 seasonally adjusted for northern US heat pump

- EI = emissions intensity of electricity ≈ 200-800 ton(CO2)/GWh, depending on region

Ground-source heat pumps always produce fewer greenhouse gases than air conditioners, oil furnaces, and electric heating, but natural gas furnaces may be competitive depending on the greenhouse gas intensity of the local electricity supply. In countries like Canada and Russia with low emitting electricity infrastructure, a residential heat pump may save 5 tons of carbon dioxide per year relative to an oil furnace, or about as much as taking an average passenger car off the road. But in cities like Beijing or Pittsburgh that are highly reliant on coal for electricity production, a heat pump may result in 1 or 2 tons more carbon dioxide emissions than a natural gas furnace. For areas not served by utility natural gas infrastructure, however, no better alternative exists.

The fluids used in closed loops may be designed to be biodegradable and non-toxic, but the refrigerant used in the heat pump cabinet and in direct exchange loops was, until recently, chlorodifluoromethane, which is an ozone depleting substance. Although harmless while contained, leaks and improper end-of-life disposal contribute to enlarging the ozone hole. For new construction, this refrigerant is being phased out in favor of the ozone-friendly but potent greenhouse gas R410A. The EcoCute water heater is an air-source heat pump that uses carbon dioxide as its working fluid instead of chlorofluorocarbons. Open loop systems (i.e. those that draw ground water as opposed to closed loop systems using a borehole heat exchanger) need to be balanced by reinjecting the spent water. This prevents aquifer depletion and the contamination of soil or surface water with brine or other compounds from underground.

Before drilling the underground geology needs to be understood, and drillers need to be prepared to seal the borehole, including preventing penetration of water between strata. The unfortunate example is a geothermal heating project in Staufen im Breisgau, Germany which seems the cause of considerable damage to historical buildings there. In 2008, the city centre was reported to have risen 12 cm, after initially sinking a few millimeters. The boring tapped a naturally pressurized aquifer, and via the borehole this water entered a layer of anhydrite, which expands when wet as it forms gypsum. The swelling will stop when the anhydrite is fully reacted, and reconstruction of the city center "is not expedient until the uplift ceases." By 2010 sealing of the borehole had not been accomplished. By 2010, some sections of town had risen by 30 cm.

Ground-source heat pump technology, like building orientation, is a natural building technique (bioclimatic building).

Economics

Ground source heat pumps are characterized by high capital costs and low operational costs compared to other HVAC systems. Their overall economic benefit depends primarily on the relative costs of electricity and fuels, which are highly variable over time and across the world. Based on recent prices, ground-source heat pumps currently have lower operational costs than any other conventional heating source almost everywhere in the world. Natural gas is the only fuel with competitive operational costs, and only in a handful of countries where it is exceptionally cheap, or where electricity is exceptionally expensive. In general, a homeowner may save anywhere from 20% to 60% annually on utilities by switching from an ordinary system to a ground-source system.

Capital costs and system lifespan have received much less study until recently, and the return on investment is highly variable. The most recent data from an analysis of 2011-2012 incentive payments in the state of Maryland showed an average cost of residential systems of $1.90 per watt, or about $26,700 for a typical (4 ton) home system. An older study found the total installed cost for a system with 10 kW (3 ton) thermal capacity for a detached rural residence in the US averaged $8000–$9000 in 1995 US dollars. More recent studies found an average cost of $14,000 in 2008 US dollars for the same size system. The US Department of Energy estimates a price of $7500 on its website, last updated in 2008. One source in Canada placed prices in the range of $30,000-$34,000 Canadian dollars. The rapid escalation in system price has been accompanied by rapid improvements in efficiency and reliability. Capital costs are known to benefit from economies of scale, particularly for open loop systems, so they are more cost-effective for larger commercial buildings and harsher climates. The initial cost can be two to five times that of a conventional heating system in most residential applications, new construction or existing. In retrofits, the cost of installation is affected by the size of living area, the home's age, insulation characteristics, the geology of the area, and location of the property. Proper duct system design and mechanical air exchange should be considered in the initial system cost.

Capital costs may be offset by government subsidies; for example, Ontario offered $7000 for residential systems installed in the 2009 fiscal year. Some electric companies offer special rates to customers who install a ground-source heat pump for heating or cooling their building. Where electrical plants have larger loads during summer months and idle capacity in the winter, this increases electrical sales during the winter months. Heat pumps also lower the load peak during the summer due to the increased efficiency of heat pumps, thereby avoiding costly construction of new power plants. For the same reasons, other utility companies have started to pay for the installation of ground-source heat pumps at customer residences. They lease the systems to their customers for a monthly fee, at a net overall saving to the customer.

The lifespan of the system is longer than conventional heating and cooling systems. Good data on system lifespan is not yet available because the technology is too recent, but many early systems are still operational today after 25–30 years with routine maintenance. Most loop fields have warranties for 25 to 50 years and are expected to last at least 50 to 200 years. Ground-source heat pumps use electricity for heating the house. The higher investment above conventional oil, propane or electric systems may be returned in energy savings in 2–10 years for residential systems in the US. If compared to natural gas systems, the payback period can be much longer or non-existent. The payback period for larger commercial systems in the US is 1–5 years, even when compared to natural gas. Additionally, because geothermal heat pumps usually have no outdoor compressors or cooling towers, the risk of vandalism is reduced or eliminated, potentially extending a system's lifespan.

Payback period for installing a ground source heat pump in a detached residence			
Country	**Payback period for replacing**		
	natural gas	**heating oil**	**electric heating**
Canada	13 years	3 years	6 years
US	12 years	5 years	4 years
Germany	net loss	8 years	2 years

Notes:

- Highly variable with energy prices.

- Government subsidies not included.

- Climate differences not evaluated.

Ground source heat pumps are recognized as one of the most efficient heating and cooling systems on the market. They are often the second-most cost effective solution in extreme climates (after co-generation), despite reductions in thermal efficiency due to ground temperature. (The ground source is warmer in climates that need strong air conditioning, and cooler in climates that need strong heating.)

Commercial systems maintenance costs in the US have historically been between $0.11 to $0.22 per m² per year in 1996 dollars, much less than the average $0.54 per m² per year for conventional HVAC systems.

Governments that promote renewable energy will likely offer incentives for the consumer (residential), or industrial markets. For example, in the United States, incentives are offered both on the state and federal levels of government. In the United Kingdom the Renewable Heat Incentive provides a financial incentive for generation of renewable heat based on metered readings on an annual basis for 20 years for commercial buildings. The domestic Renewable Heat Incentive is due to be introduced in Spring 2014 for seven years and be based on deemed heat.

Installation

Because of the technical knowledge and equipment needed to design and size the system properly (and install the piping if heat fusion is required), a GSHP system installation requires a professional's services. Several installers have published real-time views of system performance in an online community of recent residential installations. The International Ground Source Heat Pump Association (IGSHPA), Geothermal Exchange Organization (GEO), the Canadian GeoExchange Coalition and the Ground Source Heat Pump Association maintain listings of qualified installers in the US, Canada and the UK. Furthermore, detailed analysis of Soil thermal conductivity for horizontal systems and formation thermal conductivity for vertical systems will generally result in more accurately design systems with a higher efficiency.

References

- Turcotte, D. L.; Schubert, G. (2002), Geodynamics (2 ed.), Cambridge, England, UK: Cambridge University Press, pp. 136–137, ISBN 978-0-521-66624-4.

- Tiwari, G. N.; Ghosal, M. K. Renewable Energy Resources: Basic Principles and Applications. Alpha Science Int'l Ltd., 2005 ISBN 1-84265-125-0

- Ronald DiPippo (2007). Geothermal Power Plants, Second Edition: Principles, Applications, Case Studies and Environmental Impact. Oxford: Butterworth-Heinemann. p. 159. ISBN 0-7506-8620-0.

- "Environmental Technology Verification Report" (PDF). U.S. Environmental Protection Agency. Archived from the original (PDF) on 2014-04-19. Retrieved December 3, 2015.

- Johnston, Hamish (19 July 2011). "Radioactive decay accounts for half of Earth's heat". PhysicsWorld.com. Institute of Physics. Retrieved 18 June 2013.

- Matek, Benjamin (September 2013), Geothermal Power:International Market Overview (PDF), Geothermal Energy Association, pp. 10, 11, retrieved 11 October 2013

- "Is Geothermal Energy Renewable and Sustainable", Energy Auditor: Your Headquarters For Smart Sustainable Living:, retrieved 9 August 2012

- "CENTROAMÉRICA: MERCADOS MAYORISTAS DE ELECTRICIDAD Y TRANSACCIONES EN EL MERCADO ELÉCTRICO REGIONAL, 2010" (PDF), CEPAL, retrieved 30 August 2011

- "Geothermal Technologies Program: Tennessee Energy Efficient Schools Initiative Ground Source Heat Pumps". Apps1.eere.energy.gov. 2010-03-29. Retrieved 2011-03-30.

- Khan, M. Ali (2007), The Geysers Geothermal Field, an Injection Success Story (PDF), Annual Forum of the Groundwater Protection Council, archived from the original (PDF) on 2011-07-26, retrieved 2010-01-25

Permissions

Index